SpringerBriefs in Applied Sciences and Technology

Forensic and Medical Bioinformatics

Series Editors

Amit Kumar, BioAxis DNA Research Centre Private Ltd, Hyderabad, Telangana, India

Allam Appa Rao, Hyderabad, India

More information about this subseries at http://www.springer.com/series/11910

Amit Kumar · Ajit Kumar Saxena ·
Gwo Giun (Chris) Lee ·
Amita Kashyap · G. Jyothsna

Novel Coronavirus 2019

In-silico Vaccine Design and Drug Discovery

 Springer

Amit Kumar
BioAxis DNA Research Centre Private Ltd
Hyderabad, Telangana, India

Ajit Kumar Saxena
All India Institute of Medical Sciences
Patna, Bihar, India

Gwo Giun (Chris) Lee
National Cheng Kung University
Tainan, Taiwan

Amita Kashyap
BioAxis DNA Research Centre Private Ltd
Hyderabad, Telangana, India

G. Jyothsna
BioAxis DNA Research Centre Private Ltd
Hyderabad, Telangana, India

ISSN 2191-530X   ISSN 2191-5318   (electronic)
SpringerBriefs in Applied Sciences and Technology
ISSN 2196-8845   ISSN 2196-8853   (electronic)
SpringerBriefs in Forensic and Medical Bioinformatics
ISBN 978-981-15-7917-2   ISBN 978-981-15-7918-9   (eBook)
https://doi.org/10.1007/978-981-15-7918-9

This Springer imprint is published by the registered company Springer Nature Singapore Pte Ltd.
The registered company address is: 152 Beach Road, #21-01/04 Gateway East, Singapore 189721, Singapore

Preface

Hope you, your family and friends are safe. We are passing through quite a challenging time for society, and the crisis has left no one untouched. Coronavirus—COVID-19 is a common threat and declared pandemic by the World Health Organization. It is a call to action, responsibility and time to spread precautions and stop the fear going viral. The current situation affects everyone, and it demands all of us to play our part to contain this deadliest stigma to humanity. We mourn for the bereaved families and wish speedy recovery to the individuals and their families who are still suffering.

This book brings insights on current scenario of the COVID-19 along with drug and vaccine discovery approaches being worked out globally. Unfortunately, until we compile this book, there has not been any substantial success towards the cure of the COVID and whole scientific community is battling the intellectual war against this microorganism. Though MERS and nCoV-19 belong to the same family, there are visible differences in their etiology. There is a chapter on plasma therapy approach against COVID-19 which is widely experimented by several countries, but efficacy needs to be measured and studied based on the number of cases respond positive and get cured because of this technique.

Contents of the current book are useful to scientific fraternity and research students to gain whereabouts and knowledge related to in-silico vaccine and drug discovery against coronaviruses, statistics of COVID based on real-time datasets (Indian context specially), etiology of nCoV-19, therapies and tactics being workout against COVID-19 to control and check its deadliness and crisis caused because of its lethal spread.

We thank our Friends at BioAxis DNA Research Centre (P) Ltd., Hyderabad, Scientists in the Department of Pathology, All India Institute of Medical Sciences, Patna, for working hard and very meticulously to bring the current research work to a presentable stage in a quick timeline. This book is dedicated to all medical doctors, police personals, laboratory professionals and other corona warriors working day and night to save the humanity against this microorganism which we cannot see through our naked eyes. A big thank and salute to you all.

Please take a very good care of yourself, your families and friends, stay home and keep following the Government regulations. Let us make the world more resilient together.

Hyderabad, India	Amit Kumar
Patna, India	Ajit Kumar Saxena
Tainan, Taiwan	Gwo Giun (Chris) Lee
Hyderabad, India	Amita Kashyap
Hyderabad, India	G. Jyothsna

Contents

Chapter 1
Insights of NCoV 19 and COVID19

COVID 19, a contagious respiratory disease caused by Novel Corona Virus 19 or Severe Acute Respiratory Syndrome Coronavirus 2 is a major concern of the decade. A highly contagious microbe of Coronavirus family named Novel Coronavirus 19 (NCoV19) has badly affected China, USA, Italy and other European countries, India, Brazil, Russia which spread rapidly and took the whole world into its custody. COVID 19 has provoked the world with its quick spread and uncontrollable infection. It was first identified in Wuhan city of China which later spread throughout the globe causing the major pandemic of 2019.

The first sample obtained was in Wuhan, China with a patient complaining the symptoms of pneumonia of unknown etiology. Bronchoalveolar samples were collected from the victim and processed for real time PCR (RT PCR) assay. Results of the analysis revealed the identity of the samples close to beta corona virus. Further the study was extended to sequence the whole genome of the organism using illumine and nanopore sequencing [1]. In silico analysis based on bioinformatics genomic tools revealed the identity of the organism to be related to the beta corona virus 2B lineage. Further in silico annotation of the viral genes indicated that they belong to the corona virus family exhibiting features similar to this group [2]. The alignment studies of this genome to the corona group revealed the identity to be very close (96% identity) to bat SARS-like coronavirus strain BatCov.

1.1 Symptoms and Characteristics of COVID19

The basic symptoms of the disease include Dry Cough, Fever, loss of appetite, fatigue, loss of smell and difficulty in breathing [3]. Primarily lungs are affected leading to serious alteration in the respiration process and hence a prominent difficulty in breathing is reported in most of the patients [4]. However the disease is asymptomatic in most of the cases making it further a hurdle for early screening and diagnosis. The

© The Author(s), under exclusive licence to Springer Nature Singapore Pte Ltd. 2020
A. Kumar et al., *Novel Coronavirus 2019*,
SpringerBriefs in Forensic and Medical Bioinformatics,
https://doi.org/10.1007/978-981-15-7918-9_1

major concern in the disease is its contagious nature and currently no suitable drug or vaccine could be designed against the Virus. The incubation period of the virus may range between 4 and 14 days depending upon the innate immunity of the individuals [5]. The mortality rate of the COVID was reported to be 3% across the globe [6].

1.2   Statistics of COVID19

Laying its first case reported in Wuhan city of china [7] COVID 19 infection has emerged like a forest fire and has created a global pandemic of the decade. The countries experiencing the major impact of this pandemic are China, South Korea, Italy, Iran, Japan and America. Currently India is in the 2nd stage of COVID epidemic [8] and the number of victims is exponentially increasing day by day.

The pie charts (Figs. 1.1, 1.2 and 1.3) show the statistics of COVID infection in India. It can be inferred that Maharashtra shows the highest incidence (32.1%) of COVID infections in India which is followed by Tamil Nadu (14%) and Delhi (13.7%). The Mumbai district of Maharashtra has been reported to be the district with maximum COVID cases. All the other states show a relatively lower incidence

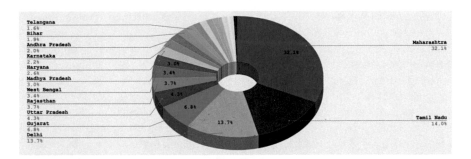

Fig. 1.1 Percentage of total cases states share

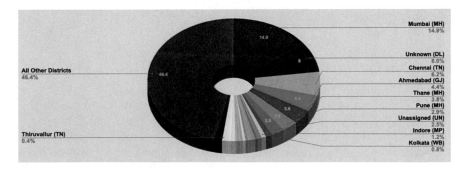

Fig. 1.2 Percentage of total cases districts share over India (Highest 30)

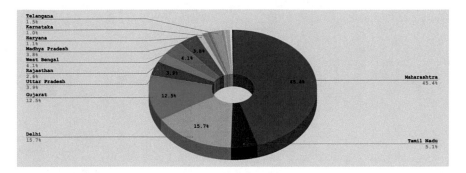

Fig. 1.3 Percentage of death cases states share

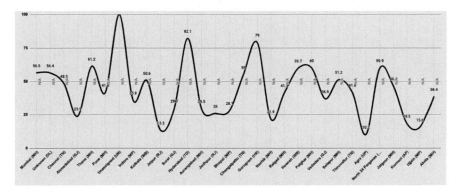

Fig. 1.4 Top 30 districts by total infected

of infection. The relative percentage of deaths due to COVID among various states of India follows the same frequency pattern-Maharashtra being the top most state in Death cases due to COVID which is followed by Tamil Nadu and Delhi.

The graphs (Figs. 1.4 and 1.5) show the COVID infection statistics and represent the top 30 districts by total infected persons as on 20th June 2020. Further, a comparison between the Mortality rate and Recovery rate of COVID infection among these 30 districts has also been done. Gurugram of Haryana shows the lowest mortality rate and also a lesser degree of recovery. The highest rate of mortality is reported in Jalgaon of Maharashtra which shows even a good recovery rate.

The graphs (Figs. 1.6, 1.7 and 1.8) show the state wise statistics of COVID Infection, Recovery rate, Death Rate, Hospitalization time etc. Both the graphs show the statistics in the form of Bubble size. Larger bubble size in the 1st Graph shows the Recovery rate of COVID Vs Death rate. Tripura shows the highest bubble size indicating a higher recovery rate. The second graph shows a comparison between Case Load Vs Period of Hospitalization.

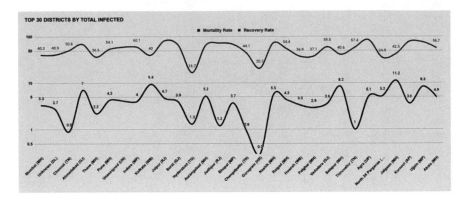

Fig. 1.5 The mortality and recovery rates in top 30 districts by total infected

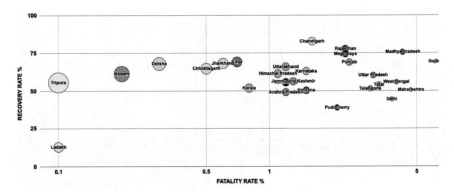

Fig. 1.6 Recovery versus death (bubble size represents recovery per death)

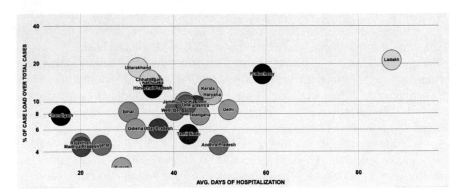

Fig. 1.7 Percentage of case load versus average period of hospitalization

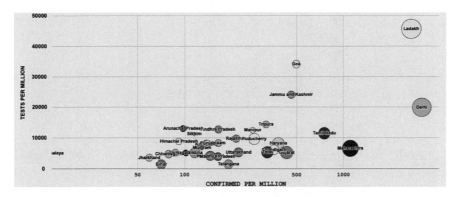

Fig. 1.8 Tests per million versus confirmed per million (bubble size represents test positivity rate). Data source of Figs. 1.1, 1.2, 1.3, 1.4, 1.5, 1.6, 1.7 and 1.8—Covid19india.org (as on 20.6.2020)

1.3 SARS Corona Virus 2

The novel corona virus is a positive sense single stranded RNA virus [9] belonging to the SARS Corona group with a little varied gene makeup making it a novel organism. The virus is spherical with numerous spikes projecting on the surface giving its peculiar appearance [10]. The genome of the organism is 27–30 KB with genes coding for different structural proteins like Membrane (M) protein, Envelope protein (E), Spike Protein (S) and Nucleocapsid (N) protein. These structural genes are essential for the virus assembly and trafficking [11]. The non-structural genes include ORF, Viral replicase etc. The spike protein of the virus enables the anchoring of these viral capsids onto the host cell membranes [12] (Fig. 1.9).

Fig. 1.9 Structure of the SARS CoV 2019 [13]. *Source* https://commons.wikimedia.org/wiki/File: 3D_medical_animation_corona_virus.jpg

1.4   World Health Organization Response towards COVID19

Most people infected with the COVID-19 virus will experience mild to moderate respiratory illness and recover without requiring special treatment. Older people, and those with underlying medical problems like cardiovascular disease, diabetes, chronic respiratory disease, and cancer are more likely to develop serious illness. The best way to prevent and slow down transmission is to be well informed about the COVID-19 virus, the disease it causes and its spread Protect yourself and others from infection by washing your hands or using an alcohol-based rub frequently and not touching your face.

The COVID-19 virus spreads primarily through droplets of saliva or discharge from the nose when an infected person coughs or sneezes, so it's important that you also practice respiratory etiquette (for example, by coughing into a flexed elbow). At this time, there are no specific vaccines or treatments for COVID-19. However, there are many ongoing clinical trials evaluating potential treatments.

One can reduce your chances of being infected or spreading COVID-19 by taking some simple precautions:

Regularly and thoroughly clean your hands with an alcohol-based hand rub or wash them with soap and water.

Maintain at least 1 m (3 ft.) distance between yourself and others.

Avoid going to crowded places.

Avoid touching eyes, nose and mouth.

Confirmed	Recovered	Deaths		
9.76M +177K	**4.92M**	**493K** +5,116		
Location		Confirmed ↓	Recovered	Deaths
United States		2.51M +44,726	771K	127K +599
Brazil		1.28M +46,860	698K	56,109 +990
Russia		628K +6,800	393K	8,969 +176
India		509K +17,296	296K	15,685 +407
United Kingdom		309K +1,380	-	43,414 +0

Fig. 1.10 Image as a screenshot from Google.com (COVID 19 spread and majorly effected countries (As on 27th June 2020, 6.00 p.m. IST)

If someone has fever, cough and difficulty breathing, seek medical attention, but call by telephone in advance if possible and follow the directions of your local health authority (Fig. 1.10).

References

1. Report of the WHO-China joint mission on coronavirus disease 2019 (COVID-19), 16–24 February 2020
2. Gundlapally J, Kumar A, Kashyap A, Saxena AK, Sanyal A (2020) In search of novel coronavirus 19 therapeutic targets. Helix 10(02):01–08. Retrieved from http://helixscientific.pub/index.php/Home/article/view/98
3. COVID-19 basics Symptoms, spread and other essential information about the new coronavirus and COVID-19
4. Watch Which organs does COVID-19 affect the most? The Hindu Net Desk, May 01, 2020
5. How high is the kill rate of coronavirus? Here's what a new study found. ET Online Last Updated: Apr 02, 2020, 04.32 PM IST
6. Smith EC, Denison MR (2013) Coronaviruses as DNA wannabes: a new model for the regulation of RNA virus replication fidelity. PLoS Pathog 9(12): e1003760, PMID: 24348241
7. Page J et al (2020) How it all started: China's early coronavirus missteps China's errors, dating back to the very first patients, were compounded by political leaders who dragged their feet to inform the public of the risks and to take decisive control measures, March 6, 2020
8. COVID-19 explainer: four stages of virus transmission, and what stage India currently finds itself in by TWC India edit team, 09 April 2020, TWC India
9. Goldsmith CS, Miller SE (2009) Modern uses of electron microscopy for detection of viruses. Clin Microbiol Rev 22(4):552–563 PMID: 19822888
10. Ruch TR, Machamer CE (2012) The coronavirus e protein: assembly and beyond. Viruses 4(3):363–382 PMID: 22590676
11. Bosch BJ et al (2003) The coronavirus spike protein is a class i virus fusion protein: structural and functional characterization of the fusion core complex. J Virol 77(16): 8801–8811, PMID: 12885899
12. https://commons.wikimedia.org/wiki/File:3D_medical_animation_corona_virus.jpg
13. Meredith S et al Coronavirus: China's Wuhan city revises death toll higher; Russia cases jump again. CNBC, Health and Science, APR 16 2020

Chapter 2
Genomics and Evolution of Novel Corona Virus 2019

NCBI is one of the world's premier websites and a master database [1] that has direct access to the genetics and protein related data of most of the known organisms. NCBI is used to retrieve the complete genome of the organism SARS Corona Virus 2019. The following is the information regarding the same:

Genome Length: 29903 bp ss-RNA

Description of the sequence: Severe acute respiratory syndrome coronavirus 2 isolate Wuhan-Hu-1, complete genome

The accession no for whole genome sequence: MN908947.3

2.1 Important Regions Within the Novel Corona Virus Genome [2]

From the data of NCBI it can be inferred that the major proteins involved in the virus structure and function are ORFs and structural proteins M, S, E and N. The longest gene region in the viral genome codes for ORF1ab followed by the gene coding for S protein. Furthermore, certain important characteristics of the genomic region are identified as follows:

- The complete genome is from 1 to 29903 bp.
- The initial 1–265 region codes for UTR un translated or leader sequence.
- The region between 266 and 21555 codes for ORF1ab polyprotein of the bacterial genes.
- 21563–25384 region codes for one of the structural proteins S of the virus called as Surface glycoprotein.
- 25393–26220 region codes for the ORF3a gene.
- E protein or Envelope protein is coded by the region 26245–26472.

© The Author(s), under exclusive licence to Springer Nature Singapore Pte Ltd. 2020
A. Kumar et al., *Novel Coronavirus 2019*,
SpringerBriefs in Forensic and Medical Bioinformatics,
https://doi.org/10.1007/978-981-15-7918-9_2

- 26523–27191 codes for Membrane Glycoprotein or M structural protein.
- ORF6 gene is coded from 27202–27387.
- 27394–27759 is coded for ORF 7a.
- 27894–28259 is coded for ORF8.
- The gene coding for Nucleocapsid phosphoprotein N lies in the region 28274–29533.
- Region from 29558–29674 codes for ORF10 protein.
- Finally the region from 29675–29903 ends as 3 prime UTR.

2.2   Linking NCoV with SARS and determining its genomic conservation

The genomic sequence of Novel Corona Virus (NCoV) was retrieved from NCBI which was subjected to BLAST run to detect the organisms sharing similarity. According to the data mining related to NCoV it was found that the organism belongs to the family of SARS which are known to cause severe acute respiratory syndrome (Fig. 2.1).

Inference: The above BLAST analysis for the complete genome of NCoV reveals its association with all the other SARS Corona virus of class 2 and share an identity of 99% to 100% with most of them.

Furthermore to understand the conservation pattern among the functional protein coding genes of NCoV genome their nucleotide sequences were retrieved from NCBI. A similarity search was performed using BLAST. Among them the major gene covering the maximum length of the genome is ORF1ab polyprotein (Gene region:

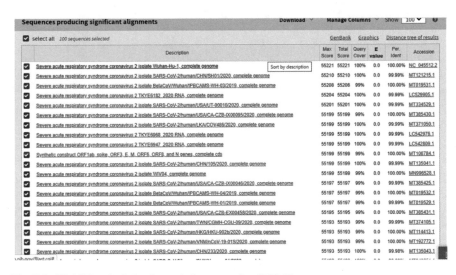

Fig. 2.1 BLAST analysis for the evolutionary study of NCoV

☑	Severe acute respiratory syndrome coronavirus 2 isolate SARS-CoV-2/human/USA/CA-CZB-EX00706/2020, complete genome	39316	39316	100%	0.0	100.00%	MT385458.1
☑	Severe acute respiratory syndrome coronavirus 2 isolate SARS-CoV-2/human/USA/CA-CZB-IX00095/2020, complete genome	39316	39316	100%	0.0	100.00%	MT385430.1
☑	Severe acute respiratory syndrome coronavirus 2 isolate SARS-CoV-2/human/CHN/Wuhan_IME-BJ05/2020, complete genome	39316	39316	100%	0.0	100.00%	MT291835.2
☑	Severe acute respiratory syndrome coronavirus 2 isolate SARS-CoV-2/human/HKG/HKU-902b/2020, complete genome	39316	39316	100%	0.0	100.00%	MT114413.1
☑	Severe acute respiratory syndrome coronavirus 2 isolate SARS-CoV-2/human/HKG/HKU-902a/2020, complete genome	39316	39316	100%	0.0	100.00%	MT114412.1
☑	Synthetic construct ORF1ab, spike, ORF3, E, M, ORF6, ORF8, and N genes, complete cds	39316	39316	100%	0.0	100.00%	MT108784.1
☑	Severe acute respiratory syndrome coronavirus 2 isolate SARS-CoV-2/human/CHN/IME-HZ01/2020, complete genome	39316	39316	100%	0.0	100.00%	MT039874.1
☑	Severe acute respiratory syndrome coronavirus 2 isolate SARS-CoV-2/human/CHN/Wuhan_IME-WH05/2019, complete genome	39316	39316	100%	0.0	100.00%	MT291830.1
☑	Severe acute respiratory syndrome coronavirus 2 isolate SARS-CoV-2/human/CHN/Wuhan_IME-WH04/2019, complete genome	39316	39316	100%	0.0	100.00%	MT291829.1
☑	Severe acute respiratory syndrome coronavirus 2 isolate SARS-CoV-2/human/CHN/Wuhan_IME-WH03/2019, complete genome	39316	39316	100%	0.0	100.00%	MT291828.1
☑	Severe acute respiratory syndrome coronavirus 2 isolate SARS-CoV-2/human/CHN/Wuhan_IME-WH02/2019, complete genome	39316	39316	100%	0.0	100.00%	MT291827.1
☑	Severe acute respiratory syndrome coronavirus 2 isolate SARS-CoV-2/human/CHN/HZ-162/2020, complete genome	39316	39316	100%	0.0	100.00%	MT253696.1
☑	Severe acute respiratory syndrome coronavirus 2 isolate nCoV-FIN-29-Jan-2020, partial genome	39316	39316	100%	0.0	100.00%	MT020781.2
☑	Severe acute respiratory syndrome coronavirus 2 isolate SARS-CoV-2/human/TWN/CGMH-CGU-01/2020, complete genome	39316	39316	100%	0.0	100.00%	MT192759.1
☑	Severe acute respiratory syndrome coronavirus 2 isolate SARS-CoV-2/human/CHN/WH-09/2020, complete genome	39316	39316	100%	0.0	100.00%	MT093631.2
☑	Severe acute respiratory syndrome coronavirus 2 isolate 2019-nCoV/USA-CruiseA-2/2020, complete genome	39316	39316	100%	0.0	100.00%	MT159718.1
☑	Severe acute respiratory syndrome coronavirus 2 isolate 2019-nCoV/USA-CruiseA-9/2020, complete genome	39316	39316	100%	0.0	100.00%	MT159710.1
☑	Severe acute respiratory syndrome coronavirus 2 isolate 2019-nCoV/USA-CruiseA-12/2020, complete genome	39316	39316	100%	0.0	100.00%	MT159709.1
☑	Severe acute respiratory syndrome coronavirus 2 isolate 2019-nCoV/USA-CA8/2020, complete genome	39316	39316	100%	0.0	100.00%	MT106053.1
☑	Severe acute respiratory syndrome coronavirus 2 isolate SARS-CoV-2/human/NPL/61-TW/2020, complete genome	39316	39316	100%	0.0	100.00%	MT072688.1
☑	Severe acute respiratory syndrome coronavirus 2 isolate HZ-1, complete genome	39316	39316	100%	0.0	100.00%	MT039873.1
☑	Severe acute respiratory syndrome coronavirus 2 isolate BetaCoV/Wuhan/IPBCAMS-WH-04/2019, complete genome	39316	39316	100%	0.0	100.00%	MT019532.1
☑	Severe acute respiratory syndrome coronavirus 2 isolate BetaCoV/Wuhan/IPBCAMS-WH-02/2019, complete genome	39316	39316	100%	0.0	100.00%	MT019530.1
☑	Wuhan seafood market pneumonia virus genome assembly, chromosome: whole_genome	39316	39316	100%	0.0	100.00%	LR757996.1
☑	Severe acute respiratory syndrome coronavirus 2 isolate WIV06, complete genome	39316	39316	100%	0.0	100.00%	MN996530.1

Fig. 2.2 BLAST analysis of ORF1ab polyprotein coding gene

266–21555 bp). The gene sequences of Surface Glycoprotein (S), Nucleocapsid Protein (N), ORF3a polyprotein, Membrane Glycoprotein (M), Envelope Protein (E), ORF6, ORF7a, ORF8 and ORF10 were the other sequences to be analyzed (Fig. 2.2).

Inference: From the above BLAST analysis of ORF1AB polyprotein gene sequence it can be inferred that all the organisms that belong to SARS Corona Virus 2 family share completely identical sequence for this gene. Thus ORF1AB is conserved evolutionarily and might not be the reason for the variations among the isolates.

A similar comparison was performed for the genes coding for **Surface Glycoprotein S, Nucleocapsid Protein N, ORF3a polyprotein, Membrane Glycoprotein M, Envelope Protein E, ORF6, ORF7a, ORF8** and **ORF10.** It was revealed that all these sequences are completely conserved and do not have any variable regions within the SARS corona 2 family. The above analysis related to the conservation study of the SARS Corona virus 2 reveals that all these 10 important genes share 99 to 100% sequence identity to the other SARS Corona Virus 2 family isolates. Thus there are no regions which are variable and are not the reason for evolution of any novel strain.

References

1. Sayers EW, Agarwala R, Bolton EE et al (2019) Database resources of the national center for biotechnology information. Nucleic Acids Res 47(D1):D23–D28. https://doi.org/10.1093/nar/gky1069
2. Wu F et al (2020) A new coronavirus associated with human respiratory disease in China. Nature 579(7798):265–269. https://doi.org/10.1038/s41586-020-2008-3 Epub 2020 Feb 3 PMID:32015508

Chapter 3
Comparing Proteomics of NCoV 19 and MERS Corona Virus

In order to develop any diagnostic or therapeutic protocol for COVID19 it is necessary to study the genome and proteome of the NCoV and annotate the proteins to identify its phylogeny and evolutionary relation with the other known organisms. The SARS NCoV and MERS Corona Virus belong to the Corona group of viruses and both are known to cause Respiratory illness. There is a very close similarity in the disease manifestation and mode of transmission in MERS and COVID forming basis for the comparison between the two organisms. Hence the proteins of novel Corona virus are compared with those of MERS Corona Virus (Fig. 3.1).

Inference: The above BLAST result shows that only 14% (**13,670–17,861**) of the Query genome can be aligned with the MERS Corona Virus genome (Middle East respiratory syndrome-related coronavirus isolate Bat-CoV/P.khulii/Italy/206645-63/2011, complete genome: **MG596803.1** [1]). However the identity in the aligned region is 70.58%. This reveals a close evolutionary relation between the two organisms.

The aligned gene region with 14% of the query gene region was from 13,670 to 17,861(4192 bp) in MERS CoV which is found to share the similarity with Novel Corona Virus 2019. This region codes for ORF1ab poly protein. Thus it can be concluded that the only region that shares the highest identity between MERS CoV and Novel Corona virus 2019 is the ORF1ab polyprotein.

© The Author(s), under exclusive licence to Springer Nature Singapore Pte Ltd. 2020   13
A. Kumar et al., *Novel Coronavirus 2019*,
SpringerBriefs in Forensic and Medical Bioinformatics,
https://doi.org/10.1007/978-981-15-7918-9_3

| Sequences producing significant alignments | | | | Download ⌄ | Manage Columns ⌄ | Show | 100 ⌄ | ❓ |

	Description	Max Score	Total Score	Query Cover	E value	Per. Ident	Accession
☑	Middle East respiratory syndrome-related coronavirus isolate Bat-CoV/Pkhuli/Italy/206645-63/2011, complete genome	767	767	14%	0.0	70.58%	MG596803.1
☑	Middle East respiratory syndrome-related coronavirus isolate Bat-CoV/H.savii/Italy/206645-40/2011, complete genome	765	765	13%	0.0	70.68%	MG596802.1
☑	Coronavirus Neoromicia/PML-PHE1/RSA/2011, complete genome	758	945	14%	0.0	70.93%	KC869678.4
☑	Middle East respiratory syndrome-related coronavirus strain Neoromicia/5038, complete genome	730	912	14%	0.0	70.83%	MF593268.1
☑	Middle East respiratory syndrome-related coronavirus isolate NL13845, complete genome	689	689	13%	0.0	70.24%	MG021451.1
☑	Middle East respiratory syndrome-related coronavirus isolate NL140422, complete genome	632	632	11%	2e-179	70.46%	MG021452.1
☑	Middle East respiratory syndrome-related coronavirus isolate NL140455, complete genome	628	628	12%	2e-178	70.30%	MG987421.1
☑	BtVs-BetaCoV/SC2013, complete genome	542	542	11%	3e-152	70.15%	KJ473821.1
☑	Middle East respiratory syndrome coronavirus isolate D389/15, complete genome	442	442	5%	3e-122	72.03%	KX108942.1
☑	Middle East respiratory syndrome coronavirus isolate camel/MERS/Ambara/118/2017, complete genome	178	178	1%	1e-42	72.65%	MK564474.1
☑	Middle East respiratory syndrome coronavirus isolate Hu/Kharj-KSA-2598/2015, complete genome	178	178	1%	1e-42	72.65%	KT806053.1
☑	Middle East respiratory syndrome-related coronavirus isolate camel/MERS/Ambara/126/2017, complete genome	172	172	1%	5e-41	72.48%	MK564475.1

Fig. 3.1 BLAST analysis of the whole genome sequence of novel corona virus with the MERS CoV genome

3.1 ORF1ab Protein Sequence comparison of NCoV and MERS CoV

MESLVPGFNEKTHVQLSLPVLQVRDVLVRGFGDSVEEVLSEARQ
HLKDGTCGLVEVEKGVLPQLEQPYVFIKRSDARTAPHGHVMVELVAELEGIQYGRSGE
TLGVLVPHVGEIPVAYRKVLLRKNGNKGAGGHSYGADLKSFDLGDELGTDPYEDFQEN
WNTKHSSGVTRELMRELNGGAYTRYVDNNFCGPDGYPLECIKDLLARAGKASCTLSEQ
LDFIDTKRGVYCCREHEHEIAWYTERSEKSYELQTPFEIKLAKKFDTFNGECPNFVFP
LNSIIKTIQPRVEKKKLDGFMGRIRSVYPVASPNECNQMCLSTLMKCDHCGETSWQTG
DFVKATCEFCGTENLTKEGATTCGYLPQNAVVKIYCPACHNSEVGPEHSLAEYHNESG
LKTILRKGGRTIAFGGCVFSYVGCHNKCAYWVPRASANIGCNHTGVVGEGSEGLNDNL
LEILQKEKVNINIVGDFKLNEEIAIILASFSASTSAFVETVKGLDYKAFKQIVESCGN
FKVTKGKAKKGAWNIGEQKSILSPLYAFASEAARVVRSIFSRTLETAQNSVRVLQKAA
ITILDGISQYSLRLIDAMMFTSDLATNNLVVMAYITGGVVQLTSQWLTNIFGTVYEKL
KPVLDWLEEKFKEGVEFLRDGWEIVKFISTCACEIVGGQIVTCAKEIKESVQTFFKLV
NKFLALCADSIIIGGAKLKALNLGETFVTHSKGLYRKCVKSREETGLLMPLKAPKEI
FLEGETLPTEVLTEEVVLKTGDLQPLEQPTSEAVEAPLVGTPVCINGLMLLEIKDTEK
YCALAPNMMVTNNTFTLKGGAPTKVTFGDDTVIEVQGYKSVNITFELDERIDKVLNEK
CSAYTVELGTEVNEFACVVADAVIKTLQPVSELLTPLGIDLDEWSMATYYLFDESGEF
KLASHMYCSFYPPDEDEEEGDCEEEEFEPSTQYEYGTEDDYQGKPLEFGATSAALQPE
EEQEEDWLDDDSQQTVGQQDGSEDNQTTTIQTIVEVQPQLEMELTPVVQTIEVNSFSG
YLKLTDNVYIKNADIVEEAKKVKPTVVVNAANVYLKHGGGVAGALNKATNNAMQVESD

DYIATNGPLKVGGSCVLSGHNLAKHCLHVVGPNVNKGEDIQLLKSAYENFNQHEVLLA
PLLSAGIFGADPIHSLRVCVDTVRTNVYLAVFDKNLYDKLVSSFLEMKSEKQVEQKIA
EIPKEEVKPFITESKPSVEQRKQDDKKIKACVEEVTTTLEETKFLTENLLLYIDINGN
LHPDSATLVSDIDITFLKKDAPYIVGDVVQEGVLTAVVIPTKKAGGTTEMLAKALRKV
PTDNYITTYPGQGLNGYTVEEAKTVLKKCKSAFYILPSIISNEKQEILGTVSWNLREM
LAHAEETRKLMPVCVETKAIVSTIQRKYKGIKIQEGVVDYGARFYFYTSKTTVASLIN
TLNDLNETLVTMPLGYVTHGLNLEEAARYMRSLKVPATVSVSSPDAVTAYNGYL
KTPEEHFIETISLAGSYKDWSYSGQSTQLGIEFLKRGDKSVYYTSNPTTFHLDGEVIT
FDNLKTLLLSLREVRTIKVFTTVDNINLHTQVVDMSMTYGQQFGPTYLDGADVTKIKPH
NSHEGKTFYVLPNDDTLRVEAFEYYHTTDPSFLGRYMSALNHTKKWKYPQVNGLTSIK
WADNNCYLATALLTLQQIELKFNPPALQDAYYRARAGEAANFCALILAYCNKTVGELG
DVRETMSYLFQHANLDSCKRVLNVVCKTCGQQQTTLKGVEAVMYMGTLSYEQFKKGVQ
IPCTCGKQATKYLVQQESPFVMMSAPPAQYELKHGTFTCASEYTGNYQCGHYKHITSK
ETLYCIDGALLTKSSEYKGPITDVFYKENSYTTTIKPVTYKLDGVVCTEIDPKLDNYY
KKDNSYFTEQPIDLVPNQPYPNASFDNFKFVCDNIKFADDLNQLTGYKKPASRELKVT
FFPDLNGDVVAIDYKHYTPSFKKGAKLLHKPIVWHVNNATNKATYKPNTWCIRCLWST
KPVETSNSFDVLKSEDAQGMDNLACEDLKPVSEEVVENPTIQKDVLECNVKTTEVVGD
IILKPANNSLKITEEVGHTDLMAAYVDNSSLTIKKPNELSRVLGLKTLATHGLAAVNS
VPWDTIANYAKPFLNKVVSTTTNIVTRCLNRVCTNYMPYFFTLLLQLCTFTRSTNSRI

KASMPTTIAKNTVKSVGKFCLEASFNYLKSPNFSKLINIIIWFLLLSVCLGSLIYSTA
ALGVLMSNLGMPSYCTGYREGYLNSTNVTIATYCTGSIPCSVCLSGLDSLDTYPSLET
IQITISSFKWDLTAFGLVAEWFLAYILFTRFFYVLGLAAIMQLFFSYFAVHFISNSWL
MWLIINLVQMAPISAMVRMYIFFASFYYVWKSYVHVVDGCNSSTCMMCYKRNRATRVE
CTTIVNGVRRSFYVYANGGKGFCKLHNWNCVNCDTFCAGSTFISDEVARDLSLQFKRP
INPTDQSSYIVDSVTVKNGSIHLYFDKAGQKTYERHSLSHFVNLDNLRANNTKGSLPI
NVIVFDGKSKCEESSAKSASVYYSQLMCQPILLLDQALVSDVGDSAEVAVKMFDAYVN
TFSSTFNVPMEKLKTLVATAEAELAKNVSLDNVLSTFISAARQGFVDSDVETKDVVEC
LKLSHQSDIEVTGDSCNNYMLTYNKVENMTPRDLGACIDCSARHINAQVAKSHNIALI
WNVKDFMSLSEQLRKQIRSAAKKNNLPFKLTCATTRQVVNVVTTKIALKGGKIVNNWL
KQLIKVTLVFLFVAAIFYLITPVHVMSKHTDFSSEIIGYKAIDGGVTRDIASTDTCFA
NKHADFDTWFSQRGGSYTNDKACPLIAAVITREVGFVVPGLPGTILRTTNGDFLHFLP
RVFSAVGNICYTPSKLIEYTDFATSACVLAAECTIFKDASGKPVPYCYDTNVLEGSVA

YESLRPDTRYVLMDGSIIQFPNTYLEGSVRVVTTFDSEYCRHGTCERSEAGVCVSTSG
RWVLNNDYYRSLPGVFCGVDAVNLLTNMFTPLIQPIGALDISASIVAGGIVAIVVTCL
AYYFMRFRRAFGEYSHVVAFNTLLFLMSFTVLCLTPVYSFLPGVYSVIYLYLTFYLTN
DVSFLAHIQWMVMFTPLVPFWITIAYIICISTKHFYWFFSNYLKRRVVFNGVSFSTFE
EAALCTFLLNKEMYLKLRSDVLLPLTQYNRYLALYNKYKYFSGAMDTTSYREAACCHL
AKALNDFSNSGSDVLYQPPQTSITSAVLQSGFRKMAFPSGKVEGCMVQVTCGTTTLNG
LWLDDVVYCPRHVICTSEDMLNPNYEDLLIRKSNHNFLVQAGNVQLRVIGHSMQNCVL
KLKVDTANPKTPKYKFVRIQPGQTFSVLACYNGSPSGVYQCAMRPNFTIKGSFLNGSC
GSVGFNIDYDCVSFCYMHHMELPTGVHAGTDLEGNFYGPFVDRQTAQAAGTDTTITVN
VLAWLYAAVINGDRWFLNRFTTTLNDFNLVAMKYNYEPLTQDHVDILGPLSAQTGIAV
LDMCASLKELLQNGMNGRTILGSALLEDEFTPFDVVRQCSGVTFQSAVKRTIKGTHHW
LLLTILTSLLVLVQSTQWSLFFFLYENAFLPFAMGIIAMSAFAMMFVKHKHAFLCLFL
LPSLATVAYFNMVYMPASWVMRIMTWLDMVDTSLSGFKLKDCVMYASAVVLLILMTAR
TVYDDGARRVWTLMNVLTLVYKVYYGNALDQAISMWALIISVTSNYSGVVTTVMFLAR
GIVFMCVEYCPIFFITGNTLQCIMLVYCFLGYFCTCYFGLFCLLNRYFRLTLGVYDYL
VSTQEFRYMNSQGLLPPKNSIDAFKLNIKLLGVGGKPCIKVATVQSKMSDVKCTSVVL
LSVLQQLRVESSSKLWAQCVQLHNDILLAKDTTEAFEKMVSLLSVLLSMQGAVDINKL
CEEMLDNRATLQAIASEFSSLPSYAAFATAQEAYEQAVANGDSEVVLKKLKKSLNVAK
SEFDRDAAMQRKLEKMADQAMTQMYKQARSEDKRAKVTSAMQTMLFTMLRKLDNDALN
NIINNARDGCVPLNIIPLTTAAKLMVVIPDYNTYKNTCDGTTFTYASALWEIQQVVDA
DSKIVQLSEISMDNSPNLAWPLIVTALRANSAVKLQNNELSPVALRQMSCAAGTTQTA
CTDDNALAYYNTTKGGRFVLALLSDLQDLKWARFPKSDGTGTIYTELEPPCRFVTDTP
KGPKVKYLYFIKGLNNLNRGMVLGSLAATVRLQAGNATEVPANSTVLSFCAFAVDAAK
AYKDYLASGGQPITNCVKMLCTHTGTGQAITVTPEANMDQESFGGASCCLYCRCHIDH
PNPKGFCDLKGKYVQIPTTCANDPVGFTLKNTVCTVCGMWKGYGCSCDQLREPMLQSA
DAQSFLNRVCGVSAARLTPCGTGTSTDVVYRAFDIYNDKVAGFAKFLKTNCCRFQEKD
EDDNLIDSYFVVKRHTFSNYQHEETIYNLLKDCPAVAKHDFFKFRIDGDMVPHISRQR
LTKYTMADLVYALRHFDEGNCDTLKEILVTYNCCDDDYFNKKDWYDFVENPDILRVYA
NLGERVRQALLKTVQFCDAMRNAGIVGVLTLDNQDLNGNWYDFGDFIQTTPGSGVPVV
DSYYSLLMPILTLTRALTAESHVDTDLTKPYIKWDLLKYDFTEERLKLFDRYFKYWDQ
TYHPNCVNCLDDRCILHCANFNVLFSTVFPPTSFGPLVRKIFVDGVPFVVSTGYHFRE
LGVVHNQDVNLHSSRLSFKELLVYAADPAMHAASGNLLLDKRTTCFSVAALTNNVAFQ
TVKPGNFNKDFYDFAVSKGFFKEGSSVELKHFFFAQDGNAAISDYDYYRYNLPTMCDI
RQLLFVVEVVDKYFDCYDGGCINANQVIVNNLDKSAGFPFNKWGKARLYYDSMSYEDQ
DALFAYTKRNVIPTITQMNLKYAISAKNRARTVAGVSICSTMTNRQFHQKLLKSIAAT
RGATVVIGTSKFYGGWHNMLKTVYSDVENPHLMGWDYPKCDRAMPNMLRIMASLVLAR

KHTTCCSLSHRFYRLANECAQVLSEMVMCGGSLYVKPGGTSSGDATTAYANSVFNICQ
AVTANVNALLSTDGNKIADKYVRNLQHRLYECLYRNRDVDTDFVNEFYAYLRKHFSMM
ILSDDAVVCFNSTYASQGLVASIKNFKSVLYYQNNVFMSEAKCWTETDLTKGPHEFCS
QHTMLVKQGDDYVYLPYPDPSRILGAGCFVDDIVKTDGTLMIERFVSLAIDAYPLTKH
PNQEYADVFHLYLQYIRKLHDELTGHMLDMYSVMLTNDNTSRYWEPEFYEAMYTPHTV
LQAVGACVLCNSQTSLRCGACIRRPFLCCKCCYDHVISTSHKLVLSVNPYVCNAPGCD
VTDVTQLYLGGMSYYCKSHKPPISFPLCANGQVFGLYKNTCVGSDNVTDFNAIATCDW
TNAGDYILANTCTERLKLFAAETLKATEETFKLSYGIATVREVLSDRELHLSWEVGKP
RPPLNRNYVFTGYRVTKNSKVQIGEYTFEKGDYGDAVVYRGTTTYKLNVGDYFVLTSH
TVMPLSAPTLVPQEHYVRITGLYPTLNISDEFSSNVANYQKVGMQKYSTLQGPPGTGK
SHFAIGLALYYPSARIVYTACSHAAVDALCEKALKYLPIDKCSRIIPARARVECFDKF
KVNSTLEQYVFCTVNALPETTADIVVFDEISMATNYDLSVVNARLRAKHYVYIGDPAQ
LPAPRTLLTKGTLEPEYFNSVCRLMKTIGPDMFLGTCRRCPAEIVDTVSALVYDNKLK
AHKDKSAQCFKMFYKGVITHDVSSAINRPQIGVVREFLTRNPAWRKAVFISPYNSQNA
VASKILGLPTQTVDSSQGSEYDYVIFTQTTETAHSCNVNRFNVAITRAKVGILCIMSD
RDLYDKLQFTSLEIPRRNVATLQAENVTGLFKDCSKVITGLHPTQAPTHLSVDTKFKT
EGLCVDIPGIPKDMTYRRLISMMGFKMNYQVNGYPNMFITREEAIRHVRAWIGFDVEG
CHATREAVGTNLPLQLGFSTGVNLVAVPTGYVDTPNNTDFSRVSAKPPPGDQFKHLIP
LMYKGLPWNVVRIKIVQMLSDTLKNLSDRVVFVLWAHGFELTSMKYFVKIGPERTCCL
CDRRATCFSTASDTYACWHHSIGFDYVYNPFMIDVQQWGFTGNLQSNHDLYCQVHGNA
HVASCDAIMTRCLAVHECFVKRVDWTIEYPIIGDELKINAACRKVQHMVVKAALLADK
FPVLHDIGNPKAIKCVPQADVEWKFYDAQPCSDKAYKIEELFYSYATHSDKFTDGVCL
FWNCNVDRYPANSIVCRFDTRVLSNLNLPGCDGGSLYVNKHAFHTPAFDKSAFVNLKQ
LPFFYYSDSPCESHGKQVVSDIDYVPLKSATCITRCNLGGAVCRHHANEYRLYLDAYN
MMISAGFSLWVYKQFDTYNLWNTFTRLQSLENVAFNVVNKGHFDGQQGEVPVSIINNT
VYTKVDGVDVELFENKTTLPVNVAFELWAKRNIKPVPEVKILNNLGVDIAANTVIWDY
KRDAPAHISTIGVCSMTDIAKKPTETICAPLTVFFDGRVDGQVDLFRNARNGVLITEG
SVKGLQPSVGPKQASLNGVTLIGEAVKTQFNYYKKVDGVVQQLPETYFTQSRNLQEFK
PRSQMEIDFLELAMDEFIERYKLEGYAFEHIVYGDFSHSQLGGLHLLIGLAKRFKESP
FELEDFIPMDSTVKNYFITDAQTGSSKCVCSVIDLLLDDFVEIIKSQDLSVVSKVVKV
TIDYTEISFMLWCKDGHVETFYPKLQSSQAWQPGVAMPNLYKMQRMLLEKCDLQNYGD
SATLPKGIMMNVAKYTQLCQYLNTLTLAVPYNMRVIHFGAGSDKGVAPGTAVLRQWLP
TGTLLVDSDLNDFVSDADSTLIGDCATVHTANKWDLIISDMYDPKTKNVTKENDSKEG
FFTYICGFIQQKLALGGSVAIKITEHSWNADLYKLMGHFAWWTAFVTNVNASSSEAFL
IGCNYLGKPREQIDGYVMHANYIFWRNTNPIQLSSYSLFDMSKFPLKLRGTAVMSLKE
GQINDMILSLLSKGRLIIRENNRVVISSDVLVNN

The above is the Proteomic sequence of ORF1AB of NCoV where, Yellow
residues-Corona Rpol N domain
Green residues-Viral Helicase
Yellow + cyan + Green residues-Region sharing identity to the MERS CoV
Proteome (Fig. 3.2).

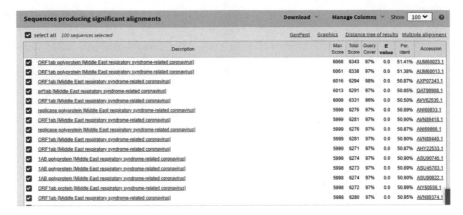

Fig. 3.2 Comparison of the ORF1ab polyprotein with the MERS CoV proteome using BLAST

Inference: The above result shows that ORF1AB protein of query (Novel Corona Virus 2019) shares 51.41% identity to that of the MERS CoV. Thus they both share a close evolutionary relation.

3.2 Comparison of Spike (S) Proteins of NCoV and MERS CoV

S protein of Novel CoV2019:

MFVFLVLLPLVSSQCVNLTTRTQLPPAYTNSFTRGVYYPDKVFR
SSVLHSTQDLFLPFFSNVTWFHAIHVSGTNGTKRFDNPVLPFNDGVYFASTEKSNIIR
GWIFGTTLDSKTQSLLIVNNATNVVIKVCEFQFCNDPFLGVYYHKNNKSWMESEFRVY
SSANNCTFEYVSQPFLMDLEGKQGNFKNLREFVFKNIDGYFKIYSKHTPINLVRDLPQ
GFSALEPLVDLPIGINITRFQTLLALHRSYLTPGDSSSGWTAGAAAYYVGYLQPRTFL
LKYNENGTITDAVDCALDPLSETKCTLKSFTVEKGIYQTSNFRVQPTESIVRFPNITN
LCPFGEVFNATRFASVYAWNRKRISNCVADYSVLYNSASFSTFKCYGVSPTKLNDLCF
TNVYADSFVIRGDEVRQIAPGQTGKIADYNYKLPDDFTGCVIAWNSNNLDSKVGGNYN
YLYRLFRKSNLKPFERDISTEIYQAGSTPCNGVEGFNCYFPLQSYGFQPTNGVGYQPY
RVVVLSFELLHAPATVCGPKKSTNLVKNKCVNFNFNGLTGTGVLTESNKKFLPFQQFG
RDIADTTDAVRDPQTLEILDITPCSFGGVSVITPGTNTSNQVAVLYQDVNCTEVPVAI
HADQLTPTWRVYSTGSNVFQTRAGCLIGAEHVNNSYECDIPIGAGICASYQTQTNSPR
RARSVASQSIIAYTMSLGAENSVAYSNNSIAIPTNFTISVTTEILPVSMTKTSVDCTM
YICGDSTECSNLLLQYGSFCTQLNRALTGIAVEQDKNTQEVFAQVKQIYKTPPIKDFG
GFNFSQILPDPSKPSKRSFIEDLLFNKVTLADAGFIKQYGDCLGDIAARDLICAQKFN
GLTVLPPLLTDEMIAQYTSALLAGTITSGWTFGAGAALQIPFAMQMAYRFNGIGVTQN
VLYENQKLIANQFNSAIGKIQDSLSSTASALGKLQDVVNQNAQALNTLVKQLSSNFGA
ISSVLNDILSRLDKVEAEVQIDRLITGRLQSLQTYVTQQLIRAAEIRASANLAATKMS
ECVLGQSKRVDFCGKGYHLMSFPQSAPHGVVFLHVTYVPAQEKNFTTAPAICHDGKAH
FPREGVFVSNGTHWFVTQRNFYEPQIITTDNTFVSGNCDVVIGIVNNTVYDPLQPELD
SFKEELDKYFKNHTSPDVDLGDISGINASVVNIQKEIDRLNEVAKNLNESLIDLQELG
KYEQYIKWPWYIWLGFIAGLIAIVMVTIMLCCMTSCCSCLKGCCSCGSCCKFDEDDSEPVLKGVKLHYT

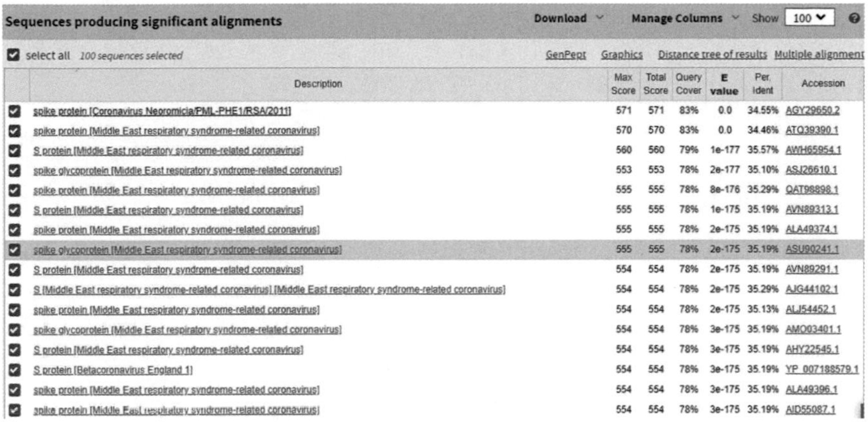

Fig. 3.3 Comparison of the S Protein with MERS CoV proteome using BLAST

Inference: From the above BLAST result it can be inferred that the S protein of Query shows an identity of about 34% to the MERS CoV Spike protein (Fig. 3.3).

3.3   Comparison of Membrane (M) Proteins of NCoV and MERS CoV

MADSNGTITVEELKKLLEQWNLVIGFLFLTWICLLQFAYANRNR
FLYIIKLIFLWLLWPVTLACFVLAAVYRINWITGGIAIAMACLVGLMWLSYFIASFRL
FARTRSMWSFNPETNILLNVPLHGTILTRPLLESELVIGAVILRGHLRIAGHHLGRCD
IKDLPKEITVATSRTLSYYKLGASQRVAGDSGFAAYSRYRIGNYKLNTDHSSSSDNIALLVQ

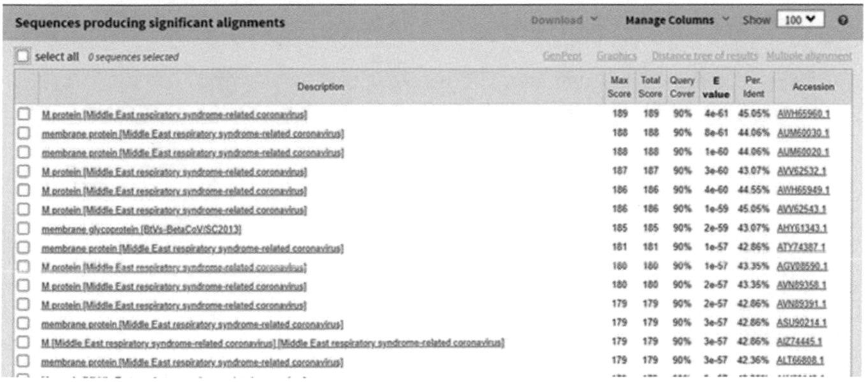

Fig. 3.4 Comparison of the M Protein with MERS CoV proteome using BLAST

Inference: The above result shows that the query M protein shares 45.05% identity to MERS CoV M protein. This can impart some minor variation in the Membrane protein structure between the two pathogens (Fig. 3.4).

3.4 Comparison of Envelope (E) Proteins of NCoV and MERS CoV

MYSFVSEETGTLIVNSVLLFLAFVVFLLVTLAILTALRLCAYCCNIVNVSLVKPSFYVYSRVKNLNSSRVPDLLV

	Description	Max Score	Total Score	Query Cover	E value	Per. Ident	Accession
☑	E protein [Middle East respiratory syndrome-related coronavirus]	51.2	51.2	94%	8e-11	40.00%	AWH65948.1
☑	small envelope protein [BtVs-BetaCoV/SC2013]	50.8	50.8	94%	1e-10	38.67%	AHY61342.1
☑	E protein [Middle East respiratory syndrome-related coronavirus]	49.7	49.7	94%	4e-10	37.33%	AVV62542.1
☑	envelope protein [Middle East respiratory syndrome-related coronavirus]	49.3	49.3	94%	4e-10	37.33%	AUM60019.1
☑	envelope protein [Middle East respiratory syndrome-related coronavirus]	48.5	48.5	94%	9e-10	37.33%	ATG39391.1
☑	E protein [Coronavirus Neoromicia/PML-PHE1/RSA/2011]	47.4	47.4	94%	3e-09	36.00%	AIG13101.1
☑	E protein [Middle East respiratory syndrome-related coronavirus]	46.6	46.6	94%	5e-09	36.00%	AVV62531.1
☑	envelope protein [Middle East respiratory syndrome-related coronavirus]	45.8	45.8	94%	1e-08	36.00%	ALA49390.1
☑	E protein [Middle East respiratory syndrome-related coronavirus]	45.1	45.1	94%	2e-08	36.00%	AGV08472.1
☑	E protein [Middle East respiratory syndrome-related coronavirus]	45.1	45.1	94%	2e-08	36.00%	AHZ65623.1
☑	E protein [Betacoronavirus England 1]	45.1	45.1	94%	2e-08	36.00%	YP_007188584.1
☑	envelope protein [Middle East respiratory syndrome-related coronavirus]	45.1	45.1	94%	2e-08	36.00%	ASU90334.1
☑	E protein [Middle East respiratory syndrome-related coronavirus]	45.1	45.1	94%	2e-08	36.00%	ANF29266.1
☑	envelope protein [Middle East respiratory syndrome-related coronavirus]	45.1	45.1	94%	2e-08	36.00%	ALR69646.1
☑	envelope [Middle East respiratory syndrome-related coronavirus]	45.1	45.1	94%	2e-08	36.00%	QBM11741.1
☑	E protein [Middle East respiratory syndrome-related coronavirus]	45.1	45.1	94%	2e-08	36.00%	ALX27237.1
☑	envelope protein [Middle East respiratory syndrome-related coronavirus]	44.7	44.7	94%	3e-08	36.00%	ANF69894.1
☑	envelope protein [Middle East respiratory syndrome-related coronavirus]	44.7	44.7	94%	3e-08	36.00%	ALA49346.1
☑	envelope protein [Middle East respiratory syndrome-related coronavirus]	44.3	44.3	94%	4e-08	36.00%	ASU90554.1
☑	E protein [Middle East respiratory syndrome-related coronavirus]	43.1	43.1	94%	1e-07	34.67%	ATG84684.1
☑	E protein [Middle East respiratory syndrome-related coronavirus]	42.7	42.7	94%	2e-07	34.67%	AGV08540.1

Fig. 3.5 Comparison of the E Protein with MERS CoV proteome using BLAST

Inference: From the above BLAST result it can be inferred that the E protein shares a considerable similarity with the E protein of MERS CoV with query coverage of 94% and identity being 40% (Fig. 3.5).

3.5 Comparison of Nucleocapsid (N) Proteins of NCoV and MERS CoV

MSDNGPQNQRNAPRITFGGPSDSTGSNQNGERSGARSKQRRPQG
LPNNTASWFTALTQHGKEDLKFPRGQGVPINTNSSPDDQIGYYRRATRRIRGGDGKMK
DLSPRWYFYYLGTGPEAGLPYGANKDGIIWVATEGALNTPKDHIGTRNPANNAAIVLQ
LPQGTTLPKGFYAEGSRGGSQASSRSSSRSRNSSRNSTPGSSRGTSPARMAGNGGDAA
LALLLLDRLNQLESKMSGKGQQQQGQTVTKKSAAEASKKPRQKRTATKAYNVTQAFGR
RGPEQTQGNFGDQELIRQGTDYKHWPQIAQFAPSASAFFGMSRIGMEVTPSGTWLTYT
GAIKLDDKDPNFKDQVILLNKHIDAYKTFPPTEPKKDKKKKADETQALPQRQKKQQTV
TLLPAADLDDFSKQLQQSMSSADSTQA

	Description	Max Score	Total Score	Query Cover	E value	Per. Ident	Accession
☑	nucleocapsid phosphoprotein [BtVs-BetaCoV/SC2013]	299	299	88%	2e-98	49.36%	AHY61344.1
☑	N protein [Middle East respiratory syndrome-related coronavirus]	288	288	88%	3e-94	50.26%	AWH65950.1
☑	N protein [Middle East respiratory syndrome-related coronavirus]	287	287	91%	7e-94	48.99%	AHY22561.1
☑	N [Middle East respiratory syndrome-related coronavirus] [Middle East respiratory syndrome-related coronavirus]	287	287	91%	7e-94	48.99%	AHL18098.1
☑	N protein [Middle East respiratory syndrome-related coronavirus]	287	287	91%	7e-94	48.99%	AHY22571.1
☑	N protein [Middle East respiratory syndrome-related coronavirus]	287	287	88%	8e-94	50.26%	AWH65961.1
☑	N protein [Middle East respiratory syndrome-related coronavirus]	286	286	91%	8e-94	48.24%	ANF29235.1
☑	N protein [Middle East respiratory syndrome-related coronavirus]	284	284	82%	2e-92	50.97%	AVV62533.1
☑	N protein [Middle East respiratory syndrome-related coronavirus]	281	281	88%	2e-91	48.97%	AVV62544.1
☑	N protein [Middle East respiratory syndrome-related coronavirus]	278	278	91%	1e-90	48.74%	ATG84774.1
☑	nucleocapsid protein [Middle East respiratory syndrome-related coronavirus]	278	278	91%	2e-90	48.74%	ASJ26617.1
☑	nucleocapsid protein [Middle East respiratory syndrome-related coronavirus]	278	278	91%	3e-90	48.49%	ALA49865.1
☑	nucleoprotein [Middle East respiratory syndrome-related coronavirus]	278	278	91%	3e-90	48.49%	AQZ41350.1
☑	nucleoprotein [Middle East respiratory syndrome-related coronavirus]	278	278	91%	3e-90	48.74%	ASU90457.1
☑	N protein [Middle East respiratory syndrome-related coronavirus]	277	277	91%	3e-90	48.29%	QBF80572.1
☑	N protein [Middle East respiratory syndrome-related coronavirus]	277	277	91%	3e-90	48.49%	ALW82676.1
☑	nucleocapsid [Middle East respiratory syndrome-related coronavirus]	277	277	91%	4e-90	48.49%	AHX00718.1
☑	nucleocapsid phosphoprotein [Middle East respiratory syndrome-related coronavirus]	277	277	91%	4e-90	48.49%	AMO03408.1
☑	nucleoprotein [Middle East respiratory syndrome-related coronavirus]	277	277	84%	4e-90	51.09%	ASU90402.1
☑	N protein [Middle East respiratory syndrome-related coronavirus]	277	277	91%	4e-90	48.49%	AVN89320.1
☑	nucleoprotein [Middle East respiratory syndrome-related coronavirus]	277	277	84%	5e-90	51.09%	ASU90380.1

select all 100 sequences selected   GenPept   Graphics   Distance tree of results   Multiple alignment

Fig. 3.6 Comparison of the N Protein with MERS CoV proteome using BLAST

Inference: The above result shows that MERS CoV Nucleocapsid shares 50.26% identity to the N protein query with coverage of 88% (Fig. 3.6).

3.6 Conclusion

From the above comparative analysis between the structural proteins and ORF1ab (being the longest protein in the virus) of NCoV and MERS Corona Virus, it can be concluded that both the organisms share a considerable similarity though not very high. However the ORF1ab is the longest protein in the Query proteome and is found to share a great similarity to that of MERS Corona Virus. Thus the study can be focused on the annotation of ORF1ab protein and targeting it for the Drug Docking studies. Further chapters of the book focus on the complete structural and functional characterization of the ORF1ab polyprotein of Novel Corona virus 2019.

References

1. Moreno A et al (2017) Detection and full genome characterization of two beta CoV viruses related to Middle East respiratory syndrome from bats in Italy. Virol J 14(1):239. https://doi.org/10.1186/s12985-017-0907-1 PMID:29258555
2. Baranov PV et al (2005) Programmed ribosomal frameshifting in decoding the SARS-CoV genome. Virology 332(2):498–510 PMID:15680415

Chapter 4
Physiochemical Characterization and Domain Annotation of ORF1ab Polyprotein of Novel Corona Virus 19

The detailed structural analysis of the ORF1ab polyprotein was carried out to study its Physiochemical characterization as well as its domain annotations. Various in silico tools have been used for the said purpose.

4.1 Physiochemical Characterization of ORF1ab Polyprotein Using Protparam

Protparam is a tool from Expasy proteomic server, for the annotation of complete physiological and chemical characteristics of the protein. The analysis is completely based on the input sequence and depends on the amino acid characteristics. A detailed annotation of the protein can be performed by the tool (Fig. 4.1).

Inference: The ORF1ab polyprotein is comprised of 7096 amino acids. The Molecular weight and the Isoelectric point are 794,057.79 and 6.32 respectively. This indicates that at the pH 6.32 the protein exists in its zwitter ionic state. Total number of negatively charged residues is the sum of Aspartic acid residues and Glutamic acid residues which is calculated to be 729 amino acids. Similarly the total number of positively charged residues is the total of Arginine and Lysine residues which is calculated to be 678. The presence of more number of acidic amino acids makes the protein acidic in nature. The instability index of the protein is 33.31 which imparts stability to the protein. The protein is hydrophilic i.e. water soluble which is indicated by its hydropathicity index being −0.070.

```
Number of amino acids: 7096

Molecular weight: 794057.79

Theoretical pI: 6.32

Amino acid composition:   CSV format
Ala (A) 487        6.9%
Arg (R) 244        3.4%
Asn (N) 384        5.4%
Asp (D) 389        5.5%
Cys (C) 226        3.2%
Gln (Q) 239        3.4%
Glu (E) 340        4.8%
Gly (G) 412        5.8%
His (H) 145        2.0%
Ile (I) 343        4.8%
Leu (L) 668        9.4%
Lys (K) 434        6.1%
Met (M) 168        2.4%
Phe (F) 349        4.9%
Pro (P) 274        3.9%
Ser (S) 456        6.4%
Thr (T) 527        7.4%
Trp (W)  78        1.1%
Tyr (Y) 335        4.7%
Val (V) 598        8.4%
Pyl (O)   0        0.0%
Sec (U)   0        0.0%

 (B)    0           0.0%
 (Z)    0           0.0%
 (X)    0           0.0%


Total number of negatively charged residues (Asp + Glu): 729
Total number of positively charged residues (Arg + Lys): 678
```

Fig. 4.1 Partial output of Protparam tool

4.2   Identification of Domains within the ORF1ab Polyprotein using SMART

Table 4.1 Domain details of the ORF1ab polyprotein

S.No.	Name	Length	Function	Sequence
1	Nsp1	115aa	Nsp1 is the N-terminal cleavage product from the viral replicase polyprotein that mediates RNA replication and processing	HVQLSLPVLQVRDVLVRGFGDSVEEVLSEAR QHLKDGTCGLVEVEKGVLPQLEQPYVFIKRS DARTAPHGHVMVELVAELEGIQYGRSGETL GVLVPHVGEIPVAYRKVLLRKNG
2	DUF3655	68AA	This domain is found in human SARS coronavirus polyprotein 1a and 1ab and in related coronavirus polyproteins. The domain, which is approximately 70 amino acids in length, identifies the N terminus	MYCSFYPPDEDEEGDCEEEFEPSTQYEYG TEDDYQGKPLEFGATSAALQPEEEQEEDWL DDDSQQT
3	A1pp	129aa	The Macro or A1pp domain is a module of about 180 amino acids which can bind ADP-ribose, an NAD metabolite or related ligands. Binding to ADP-ribose could be either covalent or non-covalent	NVYIKNADIVEEAKKVKPTVVVNAANVYLK HGGGVAGALNKATNAMQVESDDYIATNG PLKVGGSCVLSGHNLAKHCLHVVGPNVNKG EDIQLLKSAYENFNQHEVLLAPLLSAGIFGA DPIHSLRVC
4	SUDM	143aa	This domain identifies non-structural protein 3 (Nsp3), the product of ORF1a in group 2 coronavirus. It is found in human SARS coronavirus polyprotein 1a and 1ab, and in related coronavirus polyproteins	SAFYILPSIISNEKQEILGTVSWNLREMLAHA EETRKLMPVCVETKAIVSTIQRKYKGIKIQEG VVDYGARFYFYTSKTTVASLINTLNDLNETL VTMPLGYVTHGLNLEEAARYMRSLKVPATV SVSSPDAVTAYNGYLTSS
5	Nsp3_PL2pro	65aa	This domain is found in SARS and bat coronaviruses, and is about 70 amino acids in length. PL2pro is a domain of the non-structural protein nsp3, found associated with various other coronavirus proteins	PEEHFIETISLAGSYKDWSYSGQSTQLGIEFL KRGDKSVYYTSNPTTFHLDGEVITFDNLKTL LS

(continued)

Table 4.1 (continued)

S.No.	Name	Length	Function	Sequence
6	VIral Protease	319aa	This family of viral proteases is similar to the papain protease and is required for proteolytic processing of the replicase polyprotein. The structure of this protein has shown it adopts a fold	EVRTIKVFTTVDNINLHTQVVDMSMTYGQQ FGPTYLDGADVTKIKPHNSHEGKTFYVLPND DTLRVEAFEYYHTTDPSFLGRYMSALNHTK KWKYPQVNGLTSIKWADNNCYLATALLTLQ QIELKFNPPALQDAYYRARAGEAANFCALIL AYCNKTVGELGDVRETMSYLFQHANLDSCK RVLNVVCKTCGQQTTLKGVEAVMYMGTL SYEQFKKGVQIPCTCGKQATKYLVQQESPFV MMSAPPAQYELKHGTFTCASEYTGNYQCGH YKHITSKETLYCIDGALLTKSSEYKGPITDVF YKENSYTTTIKPVTY
7	NAR	111aa	This domain, approximately 100 residues in length, is mainly found in Orf1a polyproteins in severe acute respiratory syndrome coronavirus (Nucleic acid binding)	EQPIDLVPNQPYPNASFDNFKFVCDNIKFAD DLNQLTGYKKPASRELKVTFFPDLNGDVVAI DYKHYTPSFKGAKLLHKPIVWHVNNATNK ATYKPNTWCIRCLWSTKPV
8	NSP4_C	97aa	This is the C-terminal domain of the coronavirus nonstructural protein 4 (NSP4). NSP4 is a membrane-spanning protein which is thought to anchor the viral replication-transcription complex (RTC) to modified	VFNGVSFSTFEEAALCTFLLNKEMYLKLRSD VLLPLTQYNRYLALYNKYKPSGAMDTTSY REAACCHLAKALNDFSNSGSDVLYQPPQTSI TSAVL
9	Peptidase C_30	291aa	Viral protein processing. A cysteine peptidase is a proteolytic enzyme that hydrolyses a peptide bond using the thiol group of a cysteine residue as a nucleophile. Hydrolysis involves usually a catalytic triad	GLWLDDVVYCPRHVICTSEDMLNPNYEDLLI RKSNHNFLVQAGNVQLRVIGHSMQNCVLKL KVDTANPKTPKYKFVRIQPGQTFSVLACYNG SPSGVYQCAMRPNFTIKGSFLNGSCGSVGFNI DYDCVSFCYMHHMELPTGVHAGTDLEGNF YGPFVDRQTAQAAGTDTTITVNVLAWLYAA VINGDRWFLNRFTTTLNDFNLVAMKYNYEP LTQDHVDILGPLSAQTGIAVLDMCASLKELL QNGMNGRTILGSALLEDEFTPFDVVRQCSGV TFQSAVKRTIKGTHHW

Table 4.1 (continued)

S.No.	Name	Length	Function	Sequence
10	Nsp7	83aa	Nsp7 (non structural protein 7) has been implicated in viral RNA replication and is predominantly alpha helical in structure, cysteine-type endopeptidase activity (GO:0,004,197), omega peptidase activity (GO:0,008,242), transferase activity	SKMSDVKCTSVVLLSVLQQLRVESSSKLWA QCVQLHNDILLAKDTTEAFEKMVSLLSVLLS MQGAVDINKLCEEMLDNRATLQ
11	Nsp8	198aa	Viral Nsp8 (non structural protein 8) forms a hexadecameric supercomplex with Nsp7 that adopts a hollow cylinder-like structure, transferase activity (GO:0016740), cysteine-type endopeptidase activity (GO:0004197), omega peptidase activity	AIASEFSSLPSYAAFATAQEAYEQAVANGDS EVVLKKLKKSLNVAKSEFDRDAAMQRKLEK MADQAMTQMYKQARSEDKRAKVTSAMQT MLFTMLRKLDNDALNNIINNARDGCVPLNII PLTTAAKLMVVIPDYNTYKNTCDGTTFTYAS ALWEIQQVVDADSKIVQLSEISMDNSPNLAW PLIVTALRANSAVKLQ
12	Nsp9	113aa	Nsp9 is a single-stranded RNA-binding viral protein likely to be involved in RNA synthesis, viral genome replication, RNA Binding	NNELSPVALRQMSCAAGTTQTACTDDNALA YYNTTKGGRFVLALLSDLQDLKWARFPKSD GTGTIYTELEPPCRFVTDTPKGPKVKYLYFIK GLNNLNRGMVLGSLAATVRLQ
13	NSP10	123aa	Non-structural protein 10 (NSP10) is involved in RNA synthesis. It is synthesised as part of a replicase polyprotein, whose cleavage generates many non-structural proteins, viral genome replication	ANSTVLSFCAFAVDAAKAYKDYLASGGQPIT NCVKMLCTHTGTGQAITVTPEANMDQESFG GASCCLYCRCHIDHPNPKGFCDLKGKYVQIP TTCANDPVGFTLKNTVCTVCGMWKGYGC SCD

(continued)

Table 4.1 (continued)

S.No.	Name	Length	Function	Sequence
14	Corona R_pol_N	353aa	This domain represents the N-terminal region of the coronavirus RNA-directed RNA Polymerase, transcription, DNA-templated, RNA binding (GO:0003723), ATP binding (GO:0005524), RNA-directed 5′-3′ RNA polymerase activity	VSAARLTPCGTGTSTDVVYRAFDIYNDKVA GFAKFLKTNCCRFQEKDEDDNLIDSYFVVKR HTFSNYQHEETTYNLLKDCPAVAKHDFFKFR IDGDMVPHISRQRLTKYTMADLVYALRHFD EGNCDTLKEILVTYNCCDDDYFNKKDWYDF VENPDILRVYANLGERVRQALLKTVQFCDA MRNAGIVGVLTLDNQDLNGNWYDFGDFIQT TPGSGVPVVDSYYSYSLLMPILTLTRALTAESH VDTDLTKPYIKWDLLKYDFTEERLKLFDRYF KYWDQTYHPNCVNCLDDRCILHCANFNVLF STVFPPTSFGPLVRKIFVDGVPFVVSTGYHFR ELGVVHNQDVNLHSSRL
15	Viral Helicase	267aa	Helicases have been classified in 6 superfamilies (SF1–SF6) [(PUBMED:22573863)]. All of the proteins bind ATP, ATP binding	SHAAVDALCEKALKYLPIDKCSRIIPARARVE CFDKFKVNSTLEQYVFCTVNALPETTADIVV FDEISMATNYDLSVVNARLRAKHYVYIGDP AQLPAPRTLLTKGTLEPEYFNSVCRLMKTIGP DMFLGTCRRCPAEIVDTVSALVYDNKLKAH KDKSAQCFKMFYKGVITHDVSSAINRPQIGV VREFLTRNPAWRKAVFISPYNSQNAVASKIL GLPTQTVDSSOQGSEYDYVIFTQTTETAHSCN VNRFNVAITRAKVGIILCIM

(continued)

Table 4.1 (continued)

S.No.	Name	Length	Function	Sequence
16	NSP11	593aa	This region of coronavirus polyproteins encodes the NSP11 protein, cysteine-type endopeptidase activity (GO:0004197), exoribonuclease activity, producing 5'-phosphomonoesters (GO:0016896), RNA-directed 5'-3' RNA polymerase activity (GO:0003968), methyltransferase activity	NVTGLFKDCSKVITGLHPTQAPTHLSVDTKF KTEGLCVDIPGIPKDMTYRRLISMMGFKMNY QVNGYPNMFTREEAIRHVRAWIGFDVEGCH ATREAVGTNLPLQLGFSTGVNLVAVPTGYV DTPNNTDFSRVSAKPPPGDQFKHLIPLMYKG LPWNVVRIKIVQMLSDTLKNLSDRVVFVLW AHGFELTSMKYFVKIGPERTCCLCDRRATCF STASDTYACWHHSIGFDYVYNPFMIDVQQW GFTGNLQSNHDLYCQVHGNAHVASCDAIMT RCLAVHECFVKRVDWTIEYPIIGDELKINAAC RKVQHMVVKAALLADKFPVLHDIGNPKAIK CVPQADVEWKFYDAQPCSDKAYKIEELFYS YATHSDKFTDGVCLFWNCNVDRYPANSIVC RFDTRVLSNLNLPGCDGGSLYVNKHAFHTP AFDKSAFVNLKQLPFFYYSDSPCESHGKQVV SDIDYVPLKSATCITRCNLGGAVCRHHANEY RLYLDAYNMMISAGFSLWVYKQFDTYNLW NTFTRLQSLENVAFNVVNKGHFDGQQGEVP VSIINNTVYTKVDGVDVELFENKTTLPVNVA FELWAKRNIKPVPE
17	NSP13	297aa	This domain covers the NSP16 region of the coronavirus polyprotein. It was originally named NSP13 and later changed into NSP16,	SQAWQPGVAMPNLYKMQRMLLEKCDLQNY GDSATLPKGIMMNVAKYTQLCQYLNTLTLA VPYNMRVIHFGAGSDKGVAPGTAVLRQWLP TGTLLVDSDLNDFVSDADSTLIGDCATVHTA NKWDLIISDMYDPKTKNVTKENDSKEGFFTY ICGFIQQKLALGGSVAIKITEHSWNADLYKL MGHFAWWTAFVTNVNASSSEAFLIGCNYLG KPREQIDGYVMHANYIFWRNTNPIQLSSYSL FDMSKFPLKLRGTAVMSLKEGQINDMILSLL SKGRLIIRENNRVVISSDVLVNN

Fig. 4.2 Partial output of SMART

Inference: The above domain analysis as performed by SMART [1] shows that the Query sequence of ORF1ab polyprotein has a total of 17 important domains. The domains include the NSP domains which are involved in the nonstructural protein coding. The other domains include A1 Polyprotein, Viral protease, nucleic acid binding domain, Peptidase, RNA Directed RNA polymerase domain and Viral helicase domain (Fig. 4.2, Table 4.1).

4.3 The region 4406–5900 of ORF1ab polyprotein of NCoV

Further the genetic comparison of Novel Corona Virus 19 and MERS Corona Virus showed the region **4406–5900** of NCoV to be almost identical in both the groups. This region comprises of domains of evolutionary significance like Corona R pol N(4406-4758) and Viral Helicase 1(5325–5925). Thus the further study focuses on analysis of this region.

```
4406-VSAARLTPCGTGTSTDVVYRAFDIYNDKVAGFAKFLKTNCCRFQEKD
EDDNLIDSYFVVKRHTFSNYQHEETIYNLLKDCPAVAKHDFFKFRIDGDMVPHISRQR
LTKYTMADLVYALRHFDEGNCDTLKEILVTYNCCDDDYFNKKDWYDFVENPDILRVYA
NLGERVRQALLKTVQFCDAMRNAGIVGVLTLDNQDLNGNWYDFGDFIQTTPPGSGVPVV
DSYYSLLMPILTLTRALTAESHVDTDLTKPYIKWDLLKYDFTEERLKLFDRYFKYWDQ
TYHPNCVNCLDDRCILHCANFNVLFSTVFPPTSFGPLVRKIFVDGVPFVVSTGYHFRE
LGVVHNQDVNLHSSRLSFKELLVYAADPAMHAASGNLLLDKRTTCFSVAALTNNVAFQ
TVKPGNFNKDFYDFAVSKGFFKEGSSVELKHFFFAQDGNAAISDYDYYRYNLPTMCDI
RQLLFVVEVVDKYFDCYDGGCINANQVIVNNLDKSAGFPFNKWGKARLYYDSMSYEDQ
DALFAYTKRNVIPTITQMNLKYAISAKNRARTVAGVSICSTMTNRQFHQKLLKSIAAT
RGATVVIGTSKFYGGWHNMLKTVYSDVENPHLMGWDYPKCDRAMPNMLRIMASLVLAR
KHTTCCSLSHRFYRLANECAQVLSEMVMCGGSLYVKPGGTSSGDATTAYANSVFNICQ
AVTANVNALLSTDGNKIADKYVRNLQHRLYECLYRNRDVDTDFVNEFYAYLRKHFSMM
ILSDDAVVCFNSTYASQGLVASIKNFKSVLYYQNNVFMSEAKCWTETDLTKGPHEFCS
QHTMLVKQGDDYVYLPYPDPSRILGAGCFVDDIVKTDGTLMIERFVSLAIDAYPLTKH
PNQEYADVFHLYLQYIRKLHDELTGHMLDMYSVMLTNDNTSRYWEPEFYEAMYTPHTV
LQAVGACVLCNSQTSLRCGACIRRPFLCCKCCYDHVISTSHKLVLSVNPYVCNAPGCD
VTDVTQLYLGGMSYYCKSHKPPISFPLCANGQVFGLYKNTCVGSDNVTDFNAIATCDW
TNAGDYILANTCTERLKLFAAETLKATEETFKLSYGIATVREVLSDRELHLSWEVGKP
RPPLNRNYVFTGYRVTKNSKVQIGEYTFEKGDYGDAVVYRGTTTYKLNVGDYFVLTSH
TVMPLSAPTLVPQEHYVRITGLYPTLNISDEFSSNVANYQKVGMQKYSTLQGPPGTGK
SHFAIGLALYYPSARIVYTACSHAAVDALCEKALKYLPIDKCSRIIPARARVECFDKF
KVNSTLEQYVFCTVNALPETTADIVVFDEISMATNYDLSVVNARLRAKHYVYIGDPAQ
LPAPRTLLTKGTLEPEYFNSVCRLMKTIGPDMFLGTCRRCPAEIVDTVSALVYDNKLK
AHKDKSAQCFKMFYKGVITHDVSSAINRPQIGVVREFLTRNPAWRKAVFISPYNSQNA
VASKILGLPTQTVDSSQGSEYDYVIFTQTTETAHSCNVNRFNVAITRAKVGILCIM-5900
```

Length of the sequence: 1495 amino acids.
Molecular weight is: 169,186.16.
Isoelectric point: 7.31.

4.4   The Secondary Structural Conformations of the Region 4406–5900

SOPMA is an online executable tool for the structural annotation of proteins [2]. The tool predicts the detailed secondary structural confirmation of protein with respect to each amino acid in it. Based on the SOPMA annotations the composition percentage of each kind of secondary structures like helix, coil, extended strand etc. can be calculated (Fig. 4.3).

```
SOPMA :
    Alpha helix       (Hh) :    567 is   37.93%
    3₁₀ helix         (Gg) :      0 is    0.00%
    Pi helix          (Ii) :      0 is    0.00%
    Beta bridge       (Bb) :      0 is    0.00%
    Extended strand   (Ee) :    315 is   21.07%
    Beta turn         (Tt) :     99 is    6.62%
    Bend region       (Ss) :      0 is    0.00%
    Random coil       (Cc) :    514 is   34.38%
    Ambiguous states  (?)  :      0 is    0.00%
    Other states           :      0 is    0.00%
```

Fig. 4.3 Partial output of SOPMA tool

As evident, residues forming helix are more than those of sheets, confirming the hydrophilic nature of the peptide.

4.5   3D Structure prediction using Phyre

The phyre [3] result indicates that the selected sequence (region **4406–5900** of ORF1ab polyprotein of NCoV) can be structurally represented by the pdb id 7btfA. The subject sequence being represented by 7btfA shares 100% identity to the submitted query sequence with the confidence sore of 100%. The pdb id codes for the structure of SARS Corona Virus 2 RNA-dependent RNA polymerase. This structure can be used for further analysis (Fig. 4.4).

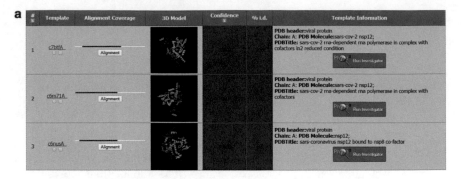

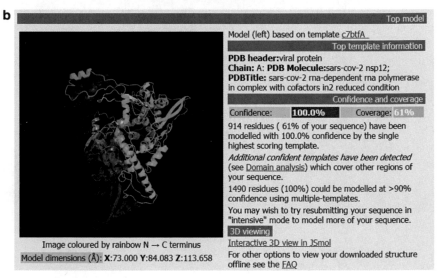

Fig. 4.4   a, **b** and **c** Partial output of Phyre and PDB

4.6   3D Structure Visualization of the SARS Corona Virus 2 RNA-Dependent RNA Polymerase Protein in RASMOL

The structure 7BTF corresponding to the Structure of SARS Corona Virus 2 RNA-Dependent RNA Polymerase was downloaded from PDB and visualized in Rasmol [4]. The subject sequence shows the structure as shown below (Fig. 4.5).

Fig. 4.5   Rasmol visualization of RNA-dependent RNA polymerase protein

SADAQSFLNRVCGVSAARLTPCGTGTSTDVVYRAFDIYNDKVAGFAKFLKTNCCRFQEKDEDDNLIDSYFVVKRH
TFSNYQHEETIYNLLKDCPAVAKHDFFKFRIDGDMVPHISRQRLTKYTMADLVYALRHFDEGNCDTLKEILVTYNC
CDDDYFNKKDWYDFVENPDILRVYANLGERVRQALLKTVQFCDAMRNAGIVGVLTLDNQDLNGNWYDFGDFIQ
TTPGSGVPVVDSYYSLLMPILTLTRALTAESHVDTDLTKPYIKWDLLKYDFTEERLKLFDRYFKYWDQTYHPNCV
NCLDDRCILHCANFNVLFSTVFPPTSFGPLVRKIFVDGVPFVVSTGYHFRELGVVHNQDVNLHSSRLSFKELLVYAA

DPAMHAASGNLLLDKRTTCFSVAALTNNVAFQTVKPGNFNKDFYDFAVSKGFFKEGSSVELKHFFFAQDGNAAIS
DYDYYRYNLPTMCDIRQLLFVVEVVDKYFDCYDGGCINANQVIVNNLDKSAGFPFNKWGKARLYYDSMSYEDQ
DALFAYTKRNVIPTITQMNLKYAISAKNRARTVAGVSICSTMTNRQFHQKLLKSIAATRGATVVIGTSKFYGGWHN
MLKTVYSDVENPHLMGWDYPKCDRAMPNMLRIMASLVLARKHTTCCSLSHRFYRLANFCAQVLSEMVMCGGSL
YVKPGGTSSGDATTAYANSVFNICQAVTANVNALLSTDGNKIADKYVRNLQHRLYECLYRNRDVDTDFVNEFYA
YLRKHFSMMILSDDAVVCFNSTYASQGLVASIKNFKSVLYYQNNVFMSEAKCWTETDLTKGPHEFCSQHTMLVK
QGGDDYVYLPYPDPSRILGAGCFVDDIVKTDGTLMIERFVSLAIDAYPLTKHPNQEYADVFHLYLQYIRKLHDELTG
HMLDMYSVMLTNDNTSRYWEPEFYEAMYTPHTVLQHHHHHHHHHH

The region that is matching between the two sequences is 14–910 of the selected gene region on the pdb structure (Fig. 4.6).

```
#========================================
#
# Aligned_sequences: 2
# 1: 7btfA
# 2: region
# Matrix: EBLOSUM62
# Gap_penalty: 10.0
# Extend_penalty: 0.5
#
# Length: 1517
# Identity:      919/1517 (60.6%)
# Similarity:    919/1517 (60.6%)
# Gaps:          597/1517 (39.4%)
# Score: 4930.0
#
#
```

Fig. 4.6 Alignment of the subject sequence with the selected protein region of 4406–5900 of query polyprotein

4.7 Prediction of Disordered Sites Within the Protein Structure Based on GLOBPLOT

Inference: The above result of Globplot [5] indicates that the regions that are more prone for mutation or the regions of disorder are 2–7, 20–29, 213–236, 321–329, 413–420, 503507, 614–624, 673–688, 828–841. The longest region within these is 213–236 (Fig. 4.7).

Fig. 4.7 Partial output of Globplot

References

1. Schultz Jörg et al (2000) SMART: a web-based tool for the study of genetically mobile domains. Nucleic Acids Res 28(1):231–234 PMID: 10592234
2. Geourjon C, Deléage G (1995) SOPMA: significant improvements in protein secondary structure prediction by consensus prediction from multiple alignments. Bioinformatics 11(6):681–684
3. Kelley L, Mezulis S, Yates C et al (2015) The Phyre2 web portal for protein modeling, prediction and analysis. Nat Protoc 10:845–858. https://doi.org/10.1038/nprot.2015.053
4. Sayle R, James Milner-White E (1995) The currently preferred literature reference to RasMol is: "RasMol: Biomolecular graphics for all". Trends in Biochem Sci (TIBS) 20(9):374
5. Linding R, Russell RB, Neduva V, Gibson TJ (2003) GlobPlot: exploring protein sequences for globularity and disorder. Nucleic Acids Res 31(13):3701–3708. https://doi.org/10.1093/nar/gkg519

Chapter 5
Rational Drug Design and Docking of the RNA Dependent RNA Polymerase Domain of NCoV

It is an urgent requirement of 2020 to develop a drug which can successfully treat the patients of COVID19 and gift them a healthy life back. The development of drugs against COVID is the need of the hour for the scientific community. However Drug development is a long term process requiring drug discovery, testing and validation and cannot be a suitable choice for the current situation. Thus instead of laying a complete focus on discovery of medication it is necessary to repurpose the available and validated drugs which were used previously against other similar virus infections. This chapter focuses on Drug repurposing and selection of suitable drug candidates which have been used previously for other similar infections and validated.

The drugs that were used in the treatment of MERS, SARS, and INFLUENZA etc. can be tried with the COVID19 victims based on their function and similarity to the NCoV. The major proteins of the virus which can be targeted in the disease are the structural proteins including Spike or S protein, Membrane or M protein, Nucleocapsid or N protein, Envelope or E protein etc. as well as the nonstructural proteins including all the NSPs from 1 to 16 and ORFs etc.

Extensive data mining works has revealed the use of ORF1ab and the Structural proteins in the treatment of MERS. Thus the same can be used in the COVID 19 therapy for the current relaxation from the virus. Some of the Drugs that can be repurposed and used for COVID19 treatment include ribavirin, glycyrrhizin and IFN-α.

SpringerBriefs in Forensic and Medical Bioinformatics,
https://doi.org/10.1007/978-981-15-7918-9_5

5.1 Selection Criteria for Chemicals

One of the important points to be considered in the treatment or the drug development for COVID 19 is time constraint. Currently Drug repurposing [1] is the best option that can be considered for the development of drugs in the treatment of the COVID victims.

In view of the above comparative analysis between MERS CoV and Novel Corona Virus 2019 which revealed a close genetic relation, all the medications that were successful in treating MERS can be used in the therapy of COVID with the molecular mechanism studied. All the drugs and their structures are obtained from two important public databases which are PUBCHEM [2] and CHEBI [3].

A **Selection of Chemicals**
1. **Remdesivir:**

This is an active inhibitor for the RNA dependent RNA polymerase enzyme which was already tested and proved to be effective against MERS [4] (https://www.jbc.org/content/early/2020/02/24/jbc.AC120.013056).

One of the study by Elfiky et al. [5] on MERS therapy stated that Guanosine derivatives are good inhibitors for virus polymerases (https://www.tandfonline.com/doi/full/10.1080/07391102.2020.1758789) and can be used in treating MERS which was validated by in silico docking studies. Thus these derivatives along with the proposed other drugs can be tested for COVID therapy. The compounds tested in the study were as follows:

2. Guanosine triphosphate (GTP)
3. Uridine triphosphate (UTP),
4. IDX-184 (GTP derivative),
5. Sofosbuvir (UTP derivative)
6. Ribavirin (wide acting antiviral drug)

B. **Collection of chemicals and their properties from Databases**:
    The data has been retrieved from two data bases CHEBI and PUBCHEM

    1. **Remdesivir [6] (CHEBI: 145994)**
       Formula: $C27H35N6O8P$
       Molecular Mass: 602.6
       Log P: 1.9
       H bond donor: 4
       H bond acceptor: 13
    2. **Guanosine triphosphate [7] (GTP) (CHEBI:16690)**
       Formula: $C10H18N5O20P5$
       Molecular Mass: 683.14
       Log P: −5.7
       H bond donor: 8
       H bond acceptors: 16

3. **Uridine triphosphate [8] (UTP) (CHEBI:15713)**
   Formula: $C_9H_{15}N_2O_{15}P_3$
   Molecular Mass: 484.14
   Log P: -5.8
   H bond donor: 7
   H bond acceptor: 15
4. **IDX-184 [9]**
   Formula: $C_{25}H_{35}N_6O_9PS$
   Molecular Mass: 626.6
   Log P: -1.3
   H bond donor: 6
   H bond acceptor: 13
5. **Sofosbuvir [10] (UTP derivative) CHEBI:85083**
   Formula: $C_{22}H_{29}FN_3O_9P$
   Molecular Mass: 529.4
   Log P: 1
   H bond donor: 3
   H bond acceptor: 11
6. **Ribavirin [11] (CHEBI:63580)**
   Formula: $C_8H_{12}N_4O_5$
   Molecular Mass: 244.2
   Log P: -1.8
   H bond donor: 4
   H bond acceptor: 7.

5.2   Docking of the Chemicals in HEX

The docking association of the above six ligands with RNA-dependent RNA polymerase domain was studied on HEX [12] and illustrated as follows:

1. **1DX-184** (Fig. 5.1).
2. **Guanosine Triphosphate** (Fig. 5.2).
3. **Uridintriphosphate** (Fig. 5.3).
4. **Sofosbuvir** (Fig. 5.4).
5. **Ribavirin** (Fig. 5.5).
6. **Remdesivir** (Fig. 5.6).

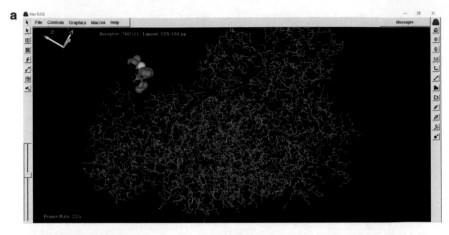

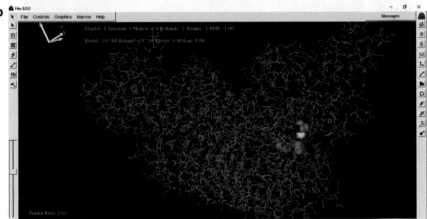

Fig. 5.1 a and **b** Different poses of 1DX-184 and RNA-dependent RNA polymerase docking at energy −297.88

5.3   Docking Results Summarized

S. No.	Ligand	HEX docking energy
1	Remdesivir	−276.31
2	Guanosine triphosphate	−253.83
3	Uridine triphosphate	−204.54
4	IDX-184	−297.88
5	Sofosbuvir	−235.87
6	Ribavirin	−162.23

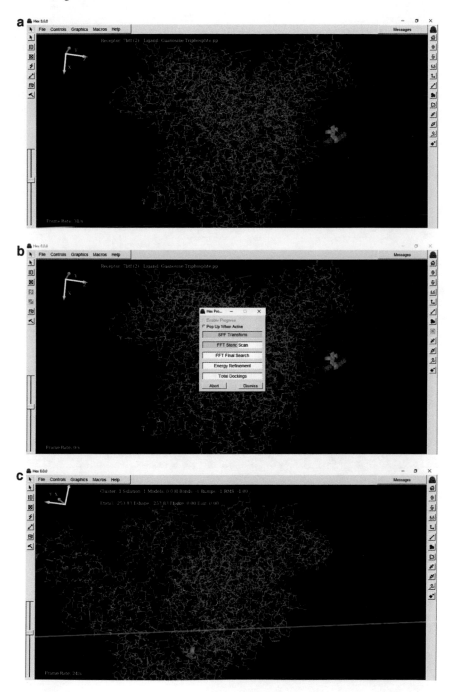

Fig. 5.2 a, **b** and **c** Different poses of Guanosine Triphosphate and RNA-dependent RNA polymerase docking at energy −253.83

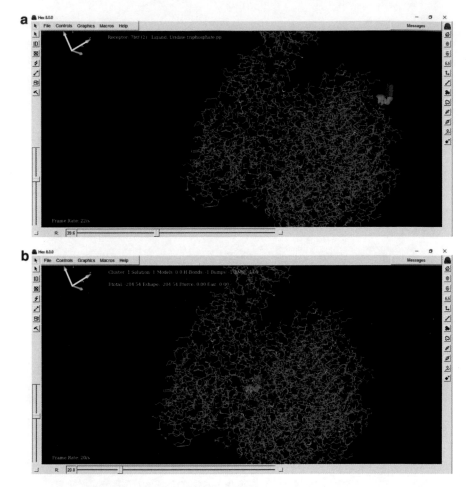

Fig. 5.3 a and **b** Different poses of Uridintriphosphate and RNA-dependent RNA polymerase docking at energy −204.54

The docking was performed between the selected ligands and the PDB structure 7btf corresponding to the RNA dependent RNA polymerase domain of ORF1AB protein. All the docking energies are energetically favorable and closer except Ribavirin. This indicates that the ligands Remdesivir, IDX-184 and Guanosine triphosphate etc can be used as efficient drugs.

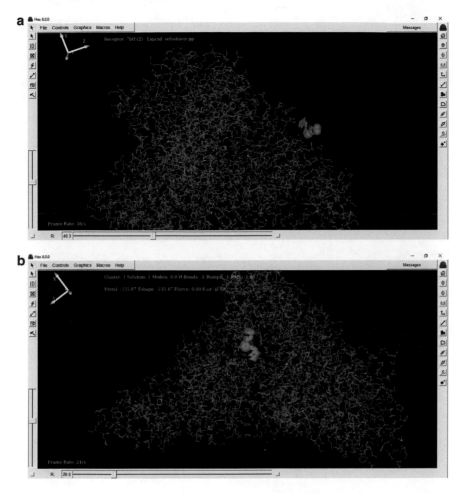

Fig. 5.4 a and **b** Different poses of Sofosbuvir and RNA-dependent RNA polymerase docking at energy −235.87

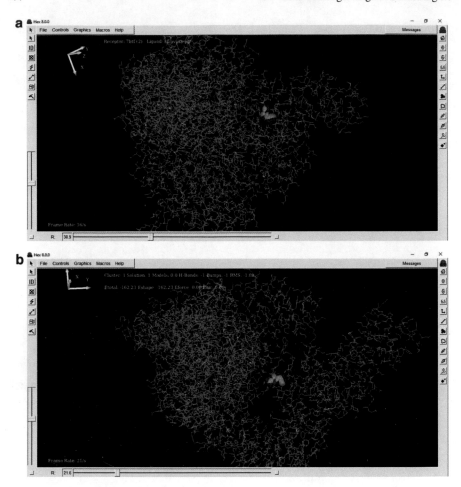

Fig. 5.5 a and **b** Different poses of Ribavirin and RNA-dependent RNA polymerase docking at energy −162.23

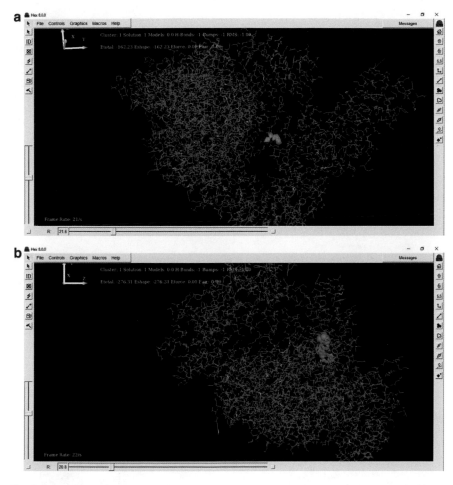

Fig. 5.6 **a** and **b** Different poses of Remdesivir and RNA-dependent RNA polymerase docking at energy −276.31

References

1. Serafin MB et al (2020) Drug repositioning is an alternative for the treatment of coronavirus COVID-19. Int J Antimicrob Agents Jun 2020; 55(6):105969. https://doi.org/10.1016/j.ijantimicag.2020.105969 Epub 2020 Apr 9
2. Kim S, Chen J, Cheng T, Gindulyte A, He J, He S, Li Q, Shoemaker BA, Thiessen PA, Yu B, Zaslavsky L, Zhang J, Bolton EE (2019) PubChem 2019 update: improved access to chemical data. Nucleic Acids Res 47(D1):D1102–D1109. https://doi.org/10.1093/nar/gky1033
3. Hastings J, de Matos P, Dekker A et al (2013) The ChEBI reference database and ontology for biologically relevant chemistry: enhancements for 2013. Nucleic Acids Res 41(Database issue):D456–D463. https://doi.org/10.1093/nar/gks1146
4. Gordon CJ et al (2020) The antiviral compound remdesivir potently inhibits RNA-dependent RNA polymerase from Middle East respiratory syndrome coronavirus. JBC Papers in Press.

Published on February 24, 2020 as Manuscript AC120.013056

5. Elfiky AA et al (2020) Novel guanosine derivatives against MERS CoV polymerase: an in silico perspective. J Biomol Struct Dyn 27
6. Wang M, Cao R, Zhang L et al (2020) Remdesivir and chloroquine effectively inhibit the recently emerged novel coronavirus (2019-nCoV) in vitro. Cell Res 30:269–271
7. Amayed P, Carlier MF, Pantaloni D (2000) Stathmin slows down guanosine diphosphate dissociation from tubulin in a phosphorylation-controlled fashion. Biochemistry 39(40):12295–12302
8. Knobloch B et al (2011) Stability and structure of mixed-ligand metal ion complexes that contain Ni^{2+}, Cu^{2+}, or Zn^{2+}, and Histamine, as well as adenosine 5'-triphosphate (ATP4-) or uridine 5'-triphosphate (UTP(4-): an intricate network of equilibria. Chemistry 17(19):5393–5403. https://doi.org/10.1002/chem.201001931 Epub 2011 Apr 4
9. https://chem.nlm.nih.gov/chemidplus/sid/1207451953
10. https://pubchem.ncbi.nlm.nih.gov/compound/Sofosbuvir#section=Literature
11. https://www.drugbank.ca/drugs/DB00811
12. Assignment DF, Argos P (1995) Knowledge-based protein secondary structure, Proteins: Struct Funct Genet 23:566–579. ftp://ftp.ebi.ac.uk/pub/software/unix/stride

Chapter 6
Herbal Treatment Approach Towards COVID19

Nature has been the enormous medical resource of active compounds which has shown promising results. Apart from the regular allopathic medication some of the medicinal plants with a potential to kill viruses can be tested and used for the therapy making it a safer, economic and less time consuming approach.

6.1 Some of the Medicinal Plants in the COVID Therapy

Previous studies related to the herbs and their antiviral properties can be made use in the treatment of COVID as an additional supplement to the regular medication. Some of the well-known herbal products include.

6.2 Zingiber Officinale/Ginger

See Fig. 6.2.

Several studies on the medicinal property of ginger have revealed its antiviral activity. Studies based on tissue culture revealed that fresh ginger have successfully acted against the Human respiratory Syncytial Viruses (HRSV) by preventing their binding to the cells of the upper respiratory tract. One of the studies by Chang et al. (2019) has showed that the use of 300 micrograms per milliliter of fresh ginger induced the release of an antiviral protein by the cells called interferon beta [1]. The antiviral efficiency of interferon alpha in covid19 is already described by Erwan Sallard et al. in June 2020 [2].

© The Author(s), under exclusive licence to Springer Nature Singapore Pte Ltd. 2020　　47
A. Kumar et al., *Novel Coronavirus 2019*,
SpringerBriefs in Forensic and Medical Bioinformatics,
https://doi.org/10.1007/978-981-15-7918-9_6

Fig. 6.1 Dos and Don'ts as the latest advisory to states. *Picture Credit* https://economictimes.ind iatimes.com/news/politics-and-nation/ayush-pushes-traditional-cure-med-council-backs-modern-drugs/articleshow/74680699.cms?from=mdr

Fig. 6.2 Zingiber Officinale/ginger

Ginger is known since long for its activity against common cold and flu, by inhibiting the viruses causing these symptoms. Ginger has proved its efficacy against several viral infections like Norwalk virus surrogate [3], human respiratory syncytial virus [4], influenza A [5], common cold [5], herpes [6], retroviral nausea and vomiting [7].

Fig. 6.3 Allium Sativum/Garlic

6.3 Allium Sativum/Garlic

See Fig. 6.3.

Garlic, a common food ingredient in Indian and Western food is known for its natural antiviral efficiency. A study by Mehrbod et al. (2009) proved the antiviral activity of Garlic against the Influenza virus [8]. The antiviral activity of garlic has also been reported in several viral infections like influenza A and B (Fenwick and Hanley [9]), cytomegalovirus (Meng et al. [10]), rhinovirus, HIV, herpes simplex virus (Tsai et al. [11]), herpes simplex virus 2 (Weber et al. [12]), viral pneumonia, and rotavirus [13].

6.4 Tinospora Cordifolia/Giloy

See Fig. 6.4.

Giloy is a common climbing shrub found growing with other plants in the fields as a weed or roadside. It is known to have several beneficial activities as explained by herbal science. It is known to possess several medicinal and therapeutic activities including its activity against the HIV [14]. Docking studies have also revealed that

Fig. 6.4 Tinospora cordifolia/Giloy

certain alkaloids extracted from *Tinospora cordifolia* can bind to the active sites of HIV 1 protease with a good affinity [15] .

6.5   Ocimum Tenuiflorum/Tulsi and Withania Somnifera/Ashwaganda

See Figs. 6.5 and 6.6.

One of the recent report released by Acharya Balakrishna (MD, Pathanjali Ayurveda) stated that Ashwagandha, Tulsi and Tinospora when used in combination would potentially inhibit the COVID 19. Their study was further supported by in silico docking studies which revealed that the phytochemicals from Ashwagandha would inhibit the ACE2 of the host, which is the major site to which the Receptor Binding Domain (RBD) of virus attaches with its spike proteins. The docking studies of this herb with ACE2 reported a very good binding efficiency. This binding would inhibit the entry of the virus into the host cells thus preventing the infection [16] (Fig. 6.7).

Fig. 6.5 Ocimum tenuiflorum/Tulsi

Fig. 6.6 Withania Somnifera/Ashwaganda

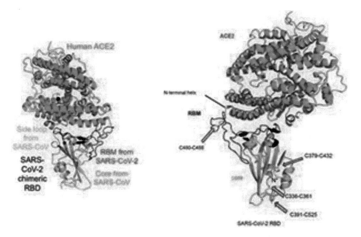

Fig. 6.7 The molecular structures of ACE2 and SARS CoV 2 RBD. *Source* © 2020 Springer Nature

Ayurveda and Herbal therapies have enough potential and possibilities to be employed both for prevention and treatment of COVID-19 [17]. This will provide an important opportunity for learning and generating credible evidence. It is pertinent to reiterate that participation of Ayurveda in addressing the COVID-19 challenge in India should not remain limited and seen as the extension of healthcare services and support to bio-medical system. Indeed, with adequate monitoring and data keeping during the implementation, important lessons and research directions are likely to emerge on the management of increasingly frequent and virulent communicable diseases [18].

However clinical observation and trials are needed for understanding efficacy of such active compounds and medicines before usages for general public. India is one of the oldest users of Ayurveda and Herbal medicines. There are 3598 AYUSH hospitals available in the country including 2818 Ayurveda hospitals. Similarly, there are 25,723 AYUSH dispensaries including 15,291 Ayurveda dispensaries [19]. There are total 7.73 lakh registered AYUSH practitioners including 4.28 lakh Ayurveda practitioners [20].

References

1. San Chang J, Wang KC, Yeh CF, Shieh DE, Chiang LC (2013) Fresh ginger (Zingiber officinale) has anti-viral activity against human respiratory syncytial virus in human respiratory tract cell lines. J Ethnopharmacol 2013 Jan 9 145(1):146–51. https://doi.org/10.1016/j.jep.2012.10.043. Epub 2012 Nov 1. https://www.ncbi.nlm.nih.gov/pubmed/23123794
2. https://www.sciencedirect.com/science/article/pii/S0166354220302059
3. https://valuecarepharmacy.net/how-fresh-ginger-can-cure-coronavirus/
4. https://www.konetou.mu/health-topics/ginger-found-to-inhibit-human-respiratory-syncytial-virus.htm

5. https://www.ncbi.nlm.nih.gov/pmc/articles/PMC5056903/
6. https://www.researchgate.net/publication/5869511_Inhibitory_effect_of_essential_oils_agai nst_Herpes_simplex_virus_type_2
7. https://www.ncbi.nlm.nih.gov/pmc/articles/PMC4818021/
8. Mehrbod P, Amini E, Tavassoti-Kheiri M (2009) Antiviral activity of garlic extract on Influenza virus. Iran J Virol 2009, 3(1):19–23. http://journal.isv.org.ir/article-1-205-en.pdf
9. Fenwick GR, Hanley AB (1985) Allium species poisoning, Vet Rec 116(1):28
10. Meng Y, Lu D, Guo N, Zhang L, Zhou G (1993) Anti-HCMV effect of garlic components. Virol Sin 8:147–150
11. Guo NL, Lu DP, Woods GL, Reed E, Zhou GZ, Zhang LB, Waldman RH (1993) Demonstration of the anti-viral activity of garlic extract against human cytomegalovirus in vitro. Chin Med J (Engl) 106(2):93–96
12. Tsai Y, Cole LL, Davis LE, Lockwood SJ, Simmons V, Wild GC (1985) Antiviral properties of garlic: in vitro effects on influenza B, herpes simplex and coxsackie viruses. Planta Med (5):460–461
13. Weber ND, Andersen DO, North JA, Murray BK, Lawson LD, Hughes BG (1992) In vitro virucidal effects of Allium sativum (garlic) extract and compounds. Planta Med 58(5):417–423
14. https://www.ncbi.nlm.nih.gov/pmc/articles/PMC3644751/
15. https://www.researchgate.net/publication/271850165_Molecular_Docking_of_HIV-1_Prot ease_using_Alkaloids_from_Tinospora_cordifolia
16. https://www.businesstoday.in/current/corporate/have-ashwagandha-giloy-tulsi-to-fight-cor onavirus-says-patanjalis-balkrishna/story/400102.html
17. Rastogi S et al (2020) COVID-19 pandemic: a pragmatic plan for Ayurveda Intervention. J Ayurveda Integr Med https://doi.org/10.1016/j.jaim.2020.04.002. 23 Apr 2020, https://doi. org/10.1016/j.jaim.2020.04.002
18. Yi Y, Lagniton PNP, Ye S, Li E, Xu R-H (2020) Covid-19: what has been learned and to be learned about the novel coronavirus disease. Int J Biol Sci 16:1753–1766. https://doi.org/10. 7150/ijbs.45134
19. Press Information Bureau, Government of India. 2020. https://pib.Gov.In/newsite/printrelease. Aspx?Relid=137509
20. http://ayush.Gov.In/sites/default/files/16%20licensed_pharmacies%202018.Pdf

Chapter 7
Evolutionary and Structural Studies of NCoV and SARS CoV-Spike proteins and their association with ACE2 Receptor

Abstract Severe acute respiratory syndrome coronavirus-2 (SARS-CoV-2)/Novel Corona Virus Disease-19 (nCOVID-19)/COVID-19 has only been discovered recently, and so our understanding of the disease epidemiology is continuously evolving. WHO has declared it a worldwide pandemic with high morbidity and significant mortality, hence it has been announced as the global health and wealth emergency. At present there is no any specific therapy available to fight against this virus, hence the drug repositioning is the most challenging to entire scientific community. The aim of this study is to determine the mutation(s) in the sequence of the spike protein, which plays a significant role in transmission from human to human. By using bioinformatics approach first we analyzed spike protein sequence of four nearest coronavirus family that include COVID-19, bat coronavirus RaTG13, pangolian coronavirus and SARS CoV, to determine phylogenetic distance between them. The homology modeling of COVID-19 spike protein has been done by iTASSER. and the protein-protein docking with human receptor ACE2 by Frodock web based docking tool showing the less binding energy of COVID-19 (-12.7 kcal/mol) in comparison with SARS CoV (10.3 kcal/mol). Further, the superimposed structure of COVID-19 and SARS CoV viruses has been performed to find the mutational site in association with human ACE2 protein. The extensive and detailed computational analyses approaches help to identify conserved region of COVID-19 and SARS CoV. Hence, our present data might help to identify potential target site and to develop antiviral drugs/vaccine to combat this pandemic.

Keywords COVID-19 · SARS CoV · ACE2 receptor · Protein docking

7.1 Introduction

Currently, the emergency has been declared by WHO due to novel human coronavirus sporadic outbreaks in different countries. The first case of novel coronavirus (nCoV-19) has been detected in China in December, 2019, where, patients presenting with viral pneumonia like symptom caused by severe acute respiratory syndrome

© The Author(s), under exclusive licence to Springer Nature Singapore Pte Ltd. 2020   53
A. Kumar et al., *Novel Coronavirus 2019*,
SpringerBriefs in Forensic and Medical Bioinformatics,
https://doi.org/10.1007/978-981-15-7918-9_7

coronavirus 2 (SARS-CoV-2), also known as coronavirus disease 2019 (COVID-19) zoonotic origins belongs to family coronaviridae, genus betacoronavirus [1]. These viruses mostly infect animals, including birds and mammals. In human, it can be transmitted from person to person [3, 4]. In humans, they generally cause mild respiratory infections, such as those observed in the high fever, dry cough with breathing issue. However, some recent human coronavirus infections have resulted in lethal endemics, which include the SARS and MERS (Middle East Respiratory Syndrome) endemics. Both of these are caused by zoonotic coronaviruses that belong to the same genus β-coronavirus within coronaviridae. SARS-CoV originated from Southern China and caused an endemic in 2003. A total of 8098 cases of SARS were reported globally, including 774 associated deaths, and an estimated case-fatality rate of approx. 15% [2]. The first case of MERS-CoV occurred in Saudi Arabia in 2012. Since then, a total of 2494 cases of infection have been reported, including 858 associated deaths, and an estimated high case-fatality rate of 34.4%. However no case of SARS-CoV infection has been reported since 2004, similarly MERS-CoV since 2012 (WHO 2011) remains undetected. Presently, COVID-19 has spread globally to about 200 countries. Globally 3,076,185 corona cases and 211,941 deaths has been recorded (April, 2020) [3, 4] and still counting is going on.

Novel Corona Viruses are very long (32 Kbp) positive sense single-stranded RNA viruses and their structure includes four main structural proteins: the spike, membrane, envelope protein, and nucleocapsid. Viral membrane protein and peptides involved in replication of viral genetic material play an integral part in virus host interaction [5]. The spike protein of coronavirus virion particles plays a significant role in the recognition of angiotensin-converting enzyme 2 (ACE2) [6, 7]. ACE2 belongs to the renin-angiotensin-aldosterone system (RAAS), which involves in regulating blood pressure, hypertension cardiovascular and renal diseases by regulating homeostasis of blood pressure, maintain electrolyte and inflammatory activities. The renin enzyme, mainly generated in the kidney and it cleaves angiotensinogen to angiotensin I (Ang I); the angiotensin-converting enzyme 2 (ACE2) cleaves Ang I to produce angiotensin II (Ang II), a key effector of the RAAS [9]. Due to alteration in ACE2, the catalytic function modulates RAAS activity, resulting in enhanced inflammation and vascular permeability observed in the pathogenesis of inflammatory lung disease [10].

Due to lack of sufficient and accurate information about this virus, the computational approach offers a method to test hypotheses of new acknowledged target site receptor with viral spike protein; hence the efficiency of viral infection is strongly dependent on virus-associated protein-protein interactions. Various metabolites are associated with protein-protein interfaces and describe the type of chemical changes occurring between ligand and target site receptor. Thus, the rationale behind this study to determine the mutation and potential drug targets to evaluate the energetic profile of the interaction between the COVID-19 spike protein and the human cell receptor ACE2.

7.2 Materials and Methods

7.2.1 Collection of Sequences

Coronavirus family spike protein sequences were retrieved from National Center for Biotechnology Information (NCBI) protein sequence data base (https://www.ncbi.nlm.nih.gov/protein) with their reference number such as COVID-19 (YP_009724390.1), Bat Coronavirus RatG13 (QHR63300.2), Pangolian coronavirus (QIQ54048.1) and SARS CoV (ACU31032.1).

7.2.2 Phylogenetic Analysis

Mega 6.0 software has been used to construct the phylogenetic tree to establish the relationship between these four coronavirus family. Alignment of the full-length coronavirus spike proteins was performed by MUSCLE with default parameters. The neighbor joining (NJ) tree was computed from the pairwise phylogenetic distance matrix creation [11].

7.2.3 Protein Structure Homology Modeling by ITASSER

I-TASSER (Iterative Threading ASSEmbly Refinement) is a hierarchical approach to protein structure and function prediction. It first identifies structural templates from the PDB by multiple threading approach LOMETS, with full-length atomic models constructed by iterative template-based fragment assembly simulations. To create structural models of the full length COVID-19 spike protein the SARS-CoV spike glycoprotein (5XLR) has been used as a template for modeling [12, 13].

7.2.4 Protein-Protein Docking

The protein–protein complexes from the predicted structure of NCoV and ACE2 human receptor were (PDB id 1R42) downloaded from protein data bank. Further, we use FRODOCK software web based user-friendly protein–protein docking server for interaction between the viral spike protein and ACE2 receptor using molecular docking [14].

7.2.5 Structural and Functional Analysis of NCoV 19 and SARS-CoV Using Human ACE2

NCoV 19 spike protein with ACE2 human binding receptor generated by FRODOCK and SARS-CoV with human ACE2 binding receptor (PDB id 3D0G) [15] were down-loaded from the Protein Data Bank. Then, binding patterns and affinity estimations for the interaction between the viral spike and ACE2 receptor were performed using Mol Star tools for web molecular graphics [16].

7.3   Results

Through bioinformatics analysis phylogenetic studies reveal that Novel CoV belongs to a group containing SARS-CoV family. The spike glycoprotein is approximately 97% similar to bat coronavirus, 90% to pangolin coronavirus and 80% closest to SARS CoV shown in Fig. 7.1.

   The 3D structures of the COVID-19 spike protein (QHD43416.1.pdb) and SARS-CoV (3D0G) interacting with the receptor binding domain (RBD) site in human ACE2 were analyzed by Frodock web based protein-protein docking tool showing (−12.7 kcal/mol) and (−10.3 kcal/mol) respectively as shown in Tables 7.1 and Fig. 7.2a, b. The interaction pattern between the viral spike proteins is quite similar in COVID-19 as well as in SARS CoV. In the case of COVID-19 total 38 amino acid

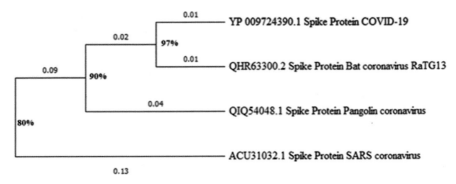

Fig. 7.1 Showing the Phylogenetic tree of Wuhan COVID-19 Spike glycoprotein sequences in context to nearest corona virus families drawn by MEGA 6.0

Table 7.1 Binding affinity (ΔG) and dissociation constant (Kd) predicted values for the interaction between viral spike and ACE2 receptor

Protein-protein complex (viral spike/ACE2)	ΔG (kcal/mol)
SARS-CoV-2	−12.7
SARS-CoV	−10.3

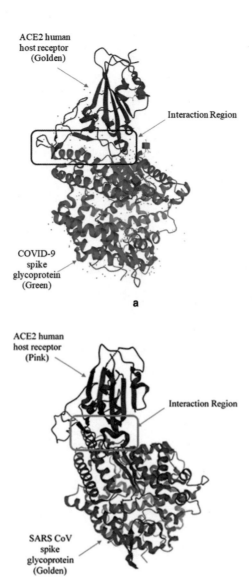

Fig. 7.2 Showing interacting binding sites **a** COVID-19 spike glycoprotein (Green) with human ACE2 receptor (Golden), **b** SARS-CoV spike glycoprotein (golden) with human ACE2 receptor (pink)

Table 7.2 Illustration of Interaction between COVID-19 and SARS-CoV spike protein with human ACE2 receptor

Protein-protein complex (viral spike/ACE2)	ΔG (kcal/mol)
SARS-CoV-2	−12.7
SARS-CoV	−10.3

residues interact with ACE2 receptor while in the case of SARS CoV total 35 amino acid residues with 22 residues of ACE2 receptor illustrated in Table 7.2.

Human ACE2 receptor COVID-19 (SARS-CoV 2) spike protein residues SARS-CoV spike protein residues

S19 A475 R426
Q24 N487 N473
T27 F456 L443, Y475
F28 Y489 Y475
D30 K417 Y442
K31 E484, F490, Q493 Y442, Y475
H34 L455, Q493 Y440, Y442, N479
E35 Q493
E37 Y505 Y491
D38 Y449, G496, Q498 Y436, G482, Y484
Y41 Q498, T500, N501 Y484, T487
Q42 G446, Y449, Q498 Y475
L45 C498, T500 T486
L79 F486 L472
M82 F486 L472
Y83 N487, Y489 N473
N330 T500 T486
K353 G496, N501 G502, Y505 Y481, G482, T487, G488, Y491
G354 G502, Y505 G488
D355 T500, G502 T487
R357 T500 G488
R393 T500 I489
Q325 R426, I489
E329 R426

In Fig. 7.3, we present a space-fill superimposed structure of spike protein showing high similarity with SARS CoV structure. The overall structure covers approx. 80% of the residues in the full-length sequence, with several important residues having in grey color, whereas red colour on superimposed structure showing mutated region and cartoon structure in cyan color representing ACE2 interaction, our following analysis based on Global Initiative on Sharing All Influenza Data (https://www.gisaid.org/hcov-19-analysis-update/) [17].

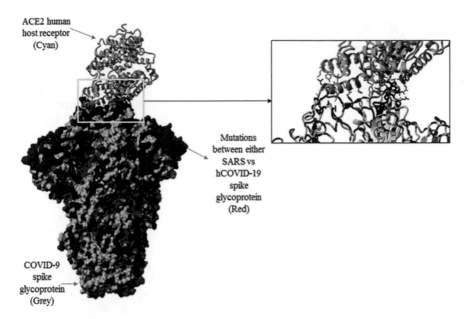

Fig. 7.3 Superimposed structure of viral spike glycoprotein of COVID-19 (grey) and SARS CoV with ACE2 (cyan) and a complex structure also showing mutational site (red) either in COVID-19 or SARS CoV [17]

7.4 Discussion

The ongoing COVID-19 pandemic makes us painfully realize that our current situation is only based on isolation from society and maintain hygienic condition to protect us from this unpredictable behavior of this virus. Earlier the outbreaks of SARS CoV in 2003 and MERS-CoV in 2012 has provide us extensive research efforts, but unfortunately there were no any drug which protect us from any type zoonotic coronavirus. Earlier, strains of these viruses were not much highly spreadable and unable to infect worldwide. But, this virus has very fluctuating and unstable strain due to their epitopic nature of the virus that make a challenges to worldwide scientific community to develop antiviral therapeutics. Due to this nature of virus earlier there was no any prototype of drug for coronavirus was progressed. After 17 years of SARS CoV epidemic and current COVID-19, emerging coronaviruses are very similar with their infection site. Therefore, our approach based on systematic comparison with these two viruses and associated mutation.

In this study, we presented a bioinformatics based methodology for systematic interaction and identification of similarity between COVID-19 and SARS CoV viral spike protein in association with ACE2 receptor binding domain. The host ACE2 has been proved by many studies to be the specific receptor for the Spike RBD of SARS-CoV [18]. The latest research shows that the host receptor of COVID-19 is consistent with SARS-CoV, exhibiting that the Spike RBD sequence of COVID-19 is similar to SARS-CoV RBD and there are important interactions between several key amino acid residues of RBD receptor-binding motif and ACE2 [8].

However, our data show the differences of interaction with considerably less favorable binding energies between these viruses with ACE2 [19], as shown in this Table 7.1. Thus, the loops observed in the spike protein of COVID-19 in the amino acids mutation (substitutions or deletion) considered as key factor with ACE2 binding [20]. Mutations in the spike protein could change the physiochemical activity of a virus which increases viral pathogenesis [21]. The presence of square bracket in (Fig. 7.2a, b) around the interaction region might be promoting the interface with ACE2 receptor, which illustrate the binding to this receptor and interaction between amino acid residues (Table 7.2).

Interestingly, mutational site could play an important clue to determine the host receptor specificity [22] for the viral spike protein which is responsible for increasing infection and viral spreading. The comparative studies to determine the impact COVID-19 and SARS CoV mutational site are quite similar with respect to receptor binding domain, which helpful and required to predict possible zoonotic event in the future as well as develop therapeutics. However, the present data might play a significant role into develop antiviral drugs and vaccines to stop the COVID-19 disease with unpredictable death.

7.5   Conclusion

Drug/vaccine development against the COVID-19 is a challenging for scientific community worldwide due to their frequent recombination events. We need explore this study on system biology to accelerate the structural and functional details of the life cycle of the COVID-19 and their mode of action. Again, as a preventive measure and strict observation of viral changes in different hosts for the prediction of an event is important aspect. Based on the current research progress, ACE2 is considered as the host potential target site for the treatment of coronavirus infection to block COVID-19 from entering host cells.

References

1. Yu T, Xia J, Wei Y, Wu W, Xie X, Yin W, Li H, Liu M, Xiao Y, Gao H, Guo L, Xie J, Wang G, Jiang R, Gao Z, Jin Q, Wang J, Cao B (2020) Clinical features of patients infected with 2019 novel coronavirus in Wuhan, China. Lancet 395:497–506
2. World-Health-Organization (2003) Update 49—SARS case fatality ratio, incubation period. Available online: https://www.who.int/csr/sars/archive/2003_05_07a/en/. Accessed 31 Jan 2020
3. World-Health-Organization (2020a) Middle East respiratory syndrome coronavirus (MERS-CoV). Available online: https://www.who.int/emergencies/mers-cov/en/. Accessed 31 Jan 2020

4. World Health Organization (WHO) (2020b) Novel Coronavirus (COVID-19) situational reports. Available online https://www.who.int/emergencies/diseases/novelcoronavirus-2020/situation-reports/98

5. Jyothsna G, Kumar A, Kashyap A, Saxena AK, Sanyal A (2020) In search of novel coronavirus 19 therapeutic targets. Helix 10(02):01–08. Retrieved from https://helixscientific.pub/index.php/Home/article/view/98

6. Zhou P, Yang X-L, Wang X-G, Hu B, Zhang L, Zhang W, Si H-R, Zhu Y, Li B, Huang C-L, Chen H-D, Chen J, Luo Y, Guo H, Jiang R-D, Liu M-Q, Chen Y, Shen X-R, Wang X, Zheng X-S, Zhao K, Chen Q-J, Deng F, Liu L-L, Yan B, Zhan F-X, Wang Y-Y, Xiao G-F, Shi Z-L (2020) A pneumonia outbreak associated with a new coronavirus of probable bat origin. Nature 579:270–273

7. Letko M, Munster V (2019) Functional assessment of cell entry and receptor usage for lineage B β-coronaviruses, including 2019-CoV. bioRxiv 2020, No. 2020.01.22.915660

8. Wan Y, Shang J, Graham R, Baric RS, Li F (2020) Receptor recognition by novel coronavirus from Wuhan: an analysis based on decade-long structural studies of SARS. J Virol 94:e00127e20

9. Jia H (2016) Pulmonary angiotensin-converting enzyme 2 (ACE2) and inflammatory lung disease. Shock 46:239–248

10. Burrell LM, Johnston CI, Tikellis C, Cooper ME (2004) ACE2, a new regulator of the renin-angiotensin system. Trends Endocrinol Metab 15:166–169

11. Edgar RC (2004) MUSCLE: multiple sequence alignment with high accuracy and high throughput. Nucleic Acids Res 32(5):1792–1797

12. Yang J, Yan R, Roy A, Xu D, Poisson J, Zhang Y (2015) The I-TASSER Suite: protein structure and function prediction. Nat Methods 12:7–8

13. Huang X, Pearcs R, Zhang Y (2020) Computational design of peptides to block binding of the SARS-CoV-2 spike protein to human ACE2. BioRxiv Prinprint https://doi.org/10.1101/2020.03.28.013607

14. Ramírez-Aportela E, López-Blanco JR, Chacón P (2016) FRODOCK 2.0: fast protein-protein docking server. Bioinformatics 32(15):2386–2388

15. Li F (2008) Structural analysis of major species barriers between humans and palm civets for severe acute respiratory syndrome coronavirus infections. J Virol 82 (14) 6984-6991

16. Sehnal D, Rose AS, Koča J, Burley SK, Velankar S (2018) Mol: towards a common library and tools for web molecular graphics. In: Proceedings of the workshop on molecular graphics and visual analysis of molecular data (MolVA'18). Eurographics Association, 2018. Goslar, DEU, 29–33

17. GISAID (2020) Genomic epidemiology of novel coronavirus. Available at https://nextstrain.org/ncov 2020 Last accessed 9 March 2020

18. Ge XY, Li JL, Yang XL, Chmura AA, Zhu G, Epstein JH et al (2013) Isolation and characterization of a bat SARS-like coronavirus that uses the ACE2 receptor. Nature 503:535e8

19. Agrawal P, Singh H, Srivastava HK et al (2019) Bench marking of different molecular docking methods for protein-peptide docking. BMC Bioinform 19:426

20. Yan R, Zhang Y, Li Y, Xia L, Guo Y, Zhou Q (2020) Structural basis for the recognition of the SARS-CoV-2 by full-length human ACE2. Science, epub ahead of print

21. Shang J, Wan Y, Liu C, Yount B, Gully K, Yang Y et al (2020) Structure of mouse coronavirus spike protein complexed with receptor reveals mechanism for viral entry. PLoS Pathog 16(3)

22. Gupta D, Kumar A (2013) Prospects for drug designing: similar conserved interactions of Bim with MCL-1 AND BCL-2

Chapter 8
Plasma Therapy Towards COVID Treatment

Apart from the drugs used in the treatment of COVID, it is necessary to try several other routes through which the recovery of the victims can be made possible. India has now laid its steps towards implementing the Plasma Therapy as an approach towards the treatment of COVID19. Here the Plasma is collected from the individuals who have recovered from the infection, as they have already developed antibodies against the pathogen. The plasma from these individuals is expected to contain antibodies against COVID which can be used further for treating the patients affected with this pathogen.

Plasma is a liquid potion of the blood that contains 90% of water and 10% of blood proteins [1]. These blood proteins mainly include antibodies, coagulation factors, serum albumin etc. The plasma of the blood can be easily separated as the supernatant by simple centrifugation of the whole blood [2]. Being highly rich in proteins and immunoglobulins plasma can be a better choice for the treatment of such diseases which do not have a perfect medication and solution. Many of the rare chronic conditions are treated with blood plasma. Plasma has been termed as a Gift of Life by some of the health organizations [3]. The concept of plasma therapy was first practically performed in the year 1982. In order to treat Diphtheria serum was collected from the animals and used for the therapy. This is a successful protocol with some mild side effects also been observed. The plasma obtained from the COVID recovered individuals who have developed the antibodies is transfused into the patients to be treated. These injected antibodies impart passive immunity to the patient.

However the chances of success are not 100% as some of the cases even revealed an elevation of infection rather than recovery. In case of dengue the use of convalescent serum has failed to provide immunity rather it had enhanced the virus replication in the host. Apart from the pathogen that is targeted in the Plasma therapy there are chances that a new pathogen may enter the victim along with plasma used for therapy causing other infections.

A. Kumar et al., *Novel Coronavirus 2019*,
SpringerBriefs in Forensic and Medical Bioinformatics,
https://doi.org/10.1007/978-981-15-7918-9_8

8.1   Assessing Global Success Rate of Plasma Therapy

Wuhan, China demonstrated successful results of Plasma Therapy where in 10 adult individuals with severe COVID Infection were subjected for plasma transfusion. Within the duration of 7 days antibodies were produced and the patients' viral load could be reduced successfully.

In case of US on March 28th 2020 a first clinical trial was conducted from the Houston Methodist Hospital where in the plasma from the blood of the COVID infected and recovered individuals was collected and transfused to the patients to be treated.

In one of the research study published in an American Journal of Pathology, the clinical outcome of the plasma transfusion for the treatment of COVID was explained. 25 patients were considered for the study among which 19 individuals exhibited a good improvement in the clinical condition with the treatment and 11 individuals could successfully be discharged from the hospital. However a concrete conclusion regarding the treatment protocol can be made only after randomized clinical trials with the large study population [4].

Recent study related to the Plasma Therapy also revealed its significance in treating previously encountered infections like SARS Pandemic in 2003, H1N1 Influenza pandemic in 2009 and most recently in 2015 the outbreak of Ebola in Africa. All these cases exhibited a good percentage of Plasma Transfusion success rate.

8.2   Plasma Therapy Guidelines and Significance in India

As a preliminary step towards the current treatment for COVID 19, India is striving to implement this system of treatment. The trials related to Plasma therapy have started in Maharashtra, Uttar Pradesh and Madhya Pradesh [5]. In particular to the Plasma therapy for COVID 19, FDA states that the donors selected for the Plasma collection should be infected 28 days prior to the plasma collection. In order to donate the plasma for the treatment the individual should be recently recovered from the COVID. There are certain parameters laid by ICMR for an individual to donate the plasma. The plasma donor should have possessed both cough and fever as symptoms during COVID infection and cannot be asymptomatic.

A recent report released by Hindustan times, Thiruvananthapuram- 17th June 2020 [6], reported a success case of Plasma Therapy in treating the COVID19 victim. Dr. MA Andrews, principal, Thrissur Medical College reports that one of the patient who returned from Delhi and facing worsened health condition was kept on ventilator support for 6 days. His worsened condition was than treated by Convalescent plasma therapy by administering the plasma obtained from the recently recovered patient of COVID. The results were successful and the patient was successfully transferred to ICU from ventilator. This was a first successful case treated with Plasma therapy in the state. This case report may be a better example to implement plasma therapy as an approach to treat COVID.

Another report released by Times of India on June 7th 2020 [7], states that Plasma therapy proved to be successful in treating a COVID patient in Kolhapur. Chatrapati Pramilaraje (CPR) Hospital has achieved success in treating a COVID patient with plasma therapy by administering the plasma containing antibodies against COVID pathogen. The plasma was obtained from the pre infected patient with COVID who has recovered in the month of April and thus developed the antibodies against Novel Corona Virus. The plasma collected is injected into the blood stream of the target patient to be treated. District collector Daulat Desai claims it to be the first successful Plasma Therapy in Kolhapur and also stated that the patient who was treated reported to be negative for COVID, thus can be a better hope for the treatment towards COVID victims.

The article also reports that Plasma Therapy is approved by the Central Government to treat critically ill and morbid patients. Also some of the recovered patients extended their interest in donating the plasma to be used for the treatment so as to maintain the plasma Stock.

Department head, PGIMER, states that the patients who do not develop fever during COVID 19 will not respond to the infection [8]. PGIMER, Chandigarh, has started the Clinical Trials for COVID and even after 3 weeks of the beginning of Clinical trials of convalescent plasma use in the patients who are critically sick, only 5 donors were selected due to the limitations and parameters proposed by ICMR. All the individuals who were asymptomatic for COVID infections are unfit for donating the plasma.

Prof Ratti Ram Sharma who is the head for the Department of Transfusion Medicine at PGIMER reported that most of the patients are asymptomatic and cannot be used for plasma donation and research. Many of the patients failed to qualify the rules of ICMR. Mohali and Jawaharpur patients were also contacted to be the donors for plasma but all the cases are asymptomatic and cannot be used. The major reason is that the patients who did not develop fever during infection fail to develop antibodies and thus cannot qualify as plasma donors (Fig. 8.1 and 8.2 and 8.3).

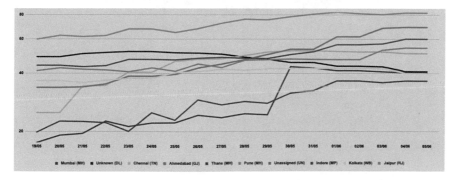

Fig. 8.1 Time series of recovery rate percentage of top ten districts by total infected

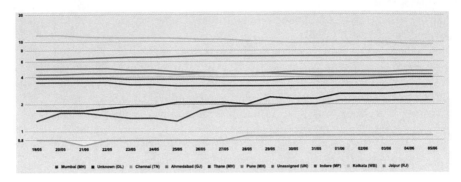

Fig. 8.2   Time series of fatality rate percentage of top ten districts by total infected

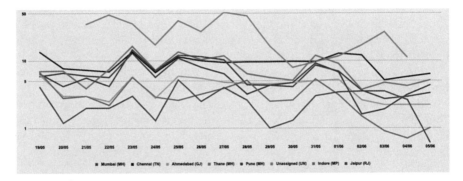

Fig. 8.3   Time series of growth rate percentage of top ten districts by total infected

References

1. COVID-19 Convalescent Plasma Program American Red Cross Blood Services, https://www.redcrossblood.org/donate-blood/dlp/plasma-information.html
2. Carmona S et al Separation of plasma from whole blood by use of the cobas plasma separation card: a compelling alternative to dried blood spots for quantification of HIV-1 viral load. J Clin Microbiol. https://doi.org/10.1128/jcm.01336-18
3. "Plasma Therapy For Covid-19", Siddha Spirituality for Health, April 27, 2020, https://siddha spirituality.com/plasma-therapy-for-covid-19/
4. New, Covid-19 plasma therapy safe, without adverse side effects: Study, Health News, Health world.com, June 3rd 2020
5. News, Covid-19: What is plasma therapy? Times of India, April 30, 2020
6. News, "Kerala: Covid-19 patient recovers after he was administered plasma therapy" Hindustan Times, June 17, 2020
7. News, "First plasma therapy for Covid-19 successful in Kolhapur", Times of India, 7 June 2020
8. News, "Asymptomatic patients pose challenge at clinical trial of plasma therapy at PGI", TributeIndia.com, May 30, 2020

Chapter 9
Identification of potential Vaccine Candidates for COVID19

The significance of vaccines in the battle of mankind with dreadful diseases cannot be surpassed. As it goes by saying 'Prevention is better than cure', vaccines have always proven their mettle. In view of the urge for the development of vaccine against the 2020 epidemic COVID the current work was undertaken. In continuation to the previous chapter on the identification of possible drug targets of COVID 19 [1] the current analysis aims to predict the possible antigenic peptides and sites on the ORF 1ab Polyprotein of Novel Corona Virus (Wuhan Isolate 2019). In view of its vital role in the replication of viral genetic material and its length covering a maximum of viral proteome, this protein is selected for the study. The protein sequence of ORF1ab poly protein was retrieved from NCBI and was subjected for BLAST analysis with Homo sapiens to identify its degree of foreignness to humans. Being a vital protein for replication it is expected to have high antigenicity and pathogenicity. It can be a better immunogen to trigger the production of antibodies by the host immune system. The study aims to develop potential vaccine candidates for stimulating the acquired immunity in the individual and prevent the adverse effects of the COVID infections.

As mentioned Corona Viruses are a group of RNA viruses [2] and pathogens of Mammals. They are further classified into various groups [3]. These are mostly winter viruses that show highest activity under low temperatures [4] however the temperature dependency is not complete. 2019–2020 can be marked as COVID era due to this major outbreak of COVID pandemic [5]. This pandemic had hit the entire world both socially and economically. Due to its contagious property it has currently become a major threat globally [6]. It has a slow activation period of 14–20 days before which the symptoms are not observed in the affected individuals. However these individuals can be good carriers for the virus even prior to its activation stage. Thus it is very difficult to control the spread of the pandemic.

Drugs can be developed to treat the infected individuals. However in order to protect the individuals from the virus vaccination is the only possible approach. Thus the current chapter involves the prediction of possible vaccine candidates/peptides against COVID using reverse vaccinology approach [7].

A. Kumar et al., *Novel Coronavirus 2019*,
SpringerBriefs in Forensic and Medical Bioinformatics,
https://doi.org/10.1007/978-981-15-7918-9_9

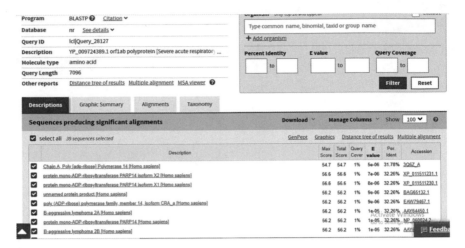

Fig. 9.1 BLAST output showing the foreignness of ORF1ab to human proteome

9.1   Determining the Foreignness of ORF1ab Poly Protein of Novel Corona Virus to Human Proteome Using BLAST

BLAST [8] is a basic local alignment search tool from NCBI. BLAST performs alignment for the comparison of any two sequences and identifies their degree of similarity. This tool is used to identify the degree of foreignness [9] between the human proteome and the targeted protein ORF1ab poly protein of Novel Corona Virus. In case the viral protein does not share any similarity, it would be recognized as FOREIGN to humans thus can be an immunogenic protein (Fig. 9.1).

Inference: The above BLAST result shows that the sequence of ORF1ab (Novel Corona Virus) shares 31% sequence similarity to Homo sapiens with a query coverage 1% only. This indicates that the sequence is not close to the human proteome and possesses foreignness; thus it can be antigenic and pathogenic and hence can be used for further analysis.

9.2   Identification of Antigenic Regions Using PVS

PVS [10] codes for Protein variability server that harbors many tools for the annotation of proteins. Antigenic Peptide prediction is a tool in PVS that would predict the antigenic regions within the protein having potential to trigger the production of antibodies in the host. Along with the antigenic peptides antigenic propensity of amino acids is also displayed in the tool output. The use of Peptide vaccines is always preferable than the whole organism Vaccine due to several parameters like less cost, high efficiency and lower risk of side effects (Fig. 9.2).

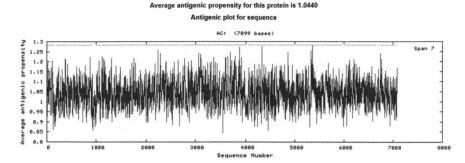

Fig. 9.2 Antigenic propensity plot of the sequence of ORF 1ab

The above graph of PVS server shows residues position number on the X axis and Antigenic propensity on Y axis. According to the graph the highest peak showing greater antigenic propensity is located between 5000 and 6000 position. This indicates that, target region to be used for vaccine peptide development lies in this region (Table 9.1).

PVS results revealed the presence of a total 311 antigenic peptides in the ORF1ab poly protein. Among them the above table highlights the regions falling under the highest peak zone. The exact peptide having highest antigenic propensity is further identified using EMBOSS ANTIGENIC tool.

Table 9.1 Details of various antigenic peptides present in the sequence of ORF 1ab

220	5153	LSDDAVVCFNS	5163
221	5165	YASQGLVAS	5173
222	5176	NFKSVLYYQN	5185
223	5204	PHEFCSQHTMLVKQGDDYVYLPYPDPSRILGAGCFVDD	5241
224	5251	IERFVSLAIDAYPLTK	5266
225	5271	EYADVFHLYLQYIRKLHD	5288
226	5295	LDMYSVM	5301
227	5320	YTPHTVLQAVGACVLCNS	5337
228	5339	TSLRCGACIRRPFLCCKCCYDHVISTSHKLVLSVNPYVCNAPGCDVTDVTQLYLGGMSYYCKSHKPPISFPLCANGQVFG	5418
229	5420	YKNTCVGS	5427
230	5445	GDYILAN	5451
231	5454	TERLKLFAAE	5463
232	5472	FKLSYGIATVREVLSDRELHLSWE	5495
233	5532	YGDAVVYR	5539
234	5544	YKLNVGDYFVLTSHTVMPLSAPTLVPQEHYVRITGLYPTL	5583
235	5590	SSNVANY	5596
236	5603	KYSTLQG	5609
237	5616	SHFAIGLALYYPSARIVYTACSHAAVDALCEKALKYLPIDKCSRIIPARARVECFDK	5672
238	5679	LEQYVFCTVNA	5689
239	5694	TADIVVFD	5701
240	5708	NYDLSVVNA	5716
241	5718	LRAKHYVYIGDPAQLPAPRT	5737
242	5748	YFNSVCRL	5755
243	5764	FLGTCRRCPAEIVDTVSALVYD	5785

9.3   Antigenic Site Prediction by EMBOSS ANTIGENIC

EMBOSS Antigenic [11] is used for the prediction of probable antigenic sites present in the above protein. The tool would predict the antigenic sites for the user entered protein sequence. The prediction is based on the Kolaskar and Tongaonkar's method. This method of antigenic site prediction is based on the physiochemical properties of the amino acids. The output gives the information about antigenic region along with the indication of site within the region with high potential for being an antigenic site. All the data obtained was tabulated.

9.4   Identification of Peptide with Highest Antigenic Propensity Using EMBOSS Antigenic Tool

See Fig. 9.3.

The above EMBOSS results show that the peptide with highest score or antigenic propensity is the region 5337–5416. This shows the highest score of 1.284 confirming

```
######################################
# Program: antigenic
# Rundate: Tue 23 Jun 2020 11:54:42
# Commandline: antigenic
#      -auto
#      -sequence /var/lib/emboss-explorer/output/199895/.sequence
#      -minlen 6
#      -outfile outfile
#      -rformat2 motif
# Report_format: motif
# Report_file: outfile
######################################

#========================================
#
# Sequence: YP_009724389.1      from: 1    to: 7096
# HitCount: 323
#========================================

Max_score_pos at "*"

(1) Score 1.284 length 80 at residues 5337->5416
                            *
  Sequence: SLRCGACIRRPFLCCKCCYDHVISTSHKLVLSVNPYVCNAPGCDVTDVTQLYLGGMSYYCKSHKPPISFPLCANGQVFGL
           |                                     |
        5337                                                                    5416
  Max_score_pos: 5352

(2) Score 1.276 length 14 at residues 4323->4336
                     *
  Sequence: GASCCLYCRCHIDH
           |            |
        4323         4336
  Max_score_pos: 4329

(3) Score 1.273 length 21 at residues 3864->3884
                     *
  Sequence: DVKCTSVVLLSVLQQLRVESS
           |                   |
```

Fig. 9.3 Partial output of EMBOSS antigenic tool

it to be most immunogenic or antigenic peptide suitable for Vaccine development. However a total of 323 peptides were predicted to be antigenic using EMBOSS.

The results of EMBOSS and PVS indicate that the following peptide is the best target for the development of peptide based vaccine due to its high degree of antigenic propensity. The K in the pattern below is the site having maximum antigenic propensity or antigenicity as identified by EMBOSS tool (Fig. 9.4).

```
SLRCGACIRRPFLCCKCCYDHVISTSHKLVLSVNPYVCNAPGCDVTDVTQLYLGGMSYYCKSHKPPISFPLCANGQVFG
L
```

Number of amino acids: 80

Molecular weight: 8764.32

Theoretical pI: 8.57

Amino acid composition: | CSV format |
```
Ala (A)    3        3.8%
Arg (R)    3        3.8%
Asn (N)    3        3.8%
Asp (D)    3        3.8%
Cys (C)   10       12.5%
Gln (Q)    2        2.5%
Glu (E)    0        0.0%
Gly (G)    6        7.5%
His (H)    3        3.8%
Ile (I)    3        3.8%
Leu (L)    8       10.0%
Lys (K)    4        5.0%
Met (M)    1        1.2%
Phe (F)    3        3.8%
Pro (P)    6        7.5%
Ser (S)    7        8.8%
Thr (T)    3        3.8%
Trp (W)    0        0.0%
Tyr (Y)    5        6.2%
Val (V)    7        8.8%
Pyl (O)    0        0.0%
Sec (U)    0        0.0%

 (B)    0        0.0%
 (Z)    0        0.0%
 (X)    0        0.0%
```

Total number of negatively charged residues (Asp + Glu): 3
Total number of positively charged residues (Arg + Lys): 7

Fig. 9.4   Partial output of protparam tool for physicochemical characterization of peptide

The above results as obtained by protparam [13] annotate the peptide in detail. The length of the peptide is found to be 80 amino acids with the molecular weight of 8764.32 and isoelectric point 8.57. This is the pH at which the peptide becomes inactive in an electric field. The peptide was identified to be unstable with the Instability index being 47.53. The peptide is basic in nature.

9.5   Construction of Peptide 3D Structure in Argus Lab

The 3D structure of peptide predicted above was constructed in Argus lab. Argus lab [12] is downloadable software for protein and chemical annotation. It is specially used for targeted docking studies and molecule and protein building. It has an inbuilt programme of energy minimization and geometry optimization of the molecule. The energy of any molecule generated in the software can be calculated.

Figure 9.5 shows the 3D structure of the selected peptide which can be used as a potential vaccine candidate. The yellow highlighted site is amino acid K (Lys) which is the one with maximum antigenic propensity. The energy of the peptide was calculated to be 2523.94 kcal/mol.

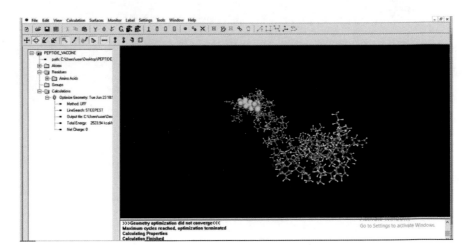

Fig. 9.5 The optimized structure of the vaccine candidate peptide designed in Argus lab

9.6 Conclusion

ORF1ab polyprotein of Novel corona virus (Wuhan Isolate 2019) was annotated further in the current research. Antigenic sites within the protein were predicted using EMBOSS ANTGENIC and were further validated by PVS (Protein variability Server). The regions showing variability are collected and subjected for further analysis. Results of the above tools were summarized to finalize a peptide with highest antigenic propensity. The peptide sequence was further annotated using Protparam to understand the physicochemical properties of the protein. Argus Lab was used to build the 3D structure of the peptide and the energy was calculated to be 2523.94 kcal/mol. The study concludes that the selected peptide is a good vaccine candidate to activate the immune response in the healthy individuals.

The work is completely based on in silico analysis, hence further laboratory study and validation is necessary before testing the efficacy of the peptide as vaccine.

References

1. Gundlapally J, Kumar A, Kashyap A, Saxena AK, Sanyal A (2020) In search of novel coronavirus 19 therapeutic targets. Helix 10(02):01–08
2. Fehr AR, Perlman S (2015) Coronaviruses: an overview of their replication and pathogenesis. Methods Mol Biol 1282:1–23. https://doi.org/10.1007/978-1-4939-2438-7_1
3. Tyrrell DAJ, Myint SH (1996) Coronaviruses. In: Baron S (ed) Medical microbiology. 4th edn. University of Texas Medical Branch at Galveston, Galveston, TX. Chapter 60. Available from: https://www.ncbi.nlm.nih.gov/books/NBK7782/
4. Razuri H, Malecki M, Tinoco Y et al (2015) Human coronavirus-associated influenza-like illness in the community setting in Peru. Am J Trop Med Hyg 93(5):1038–1040. https://doi.org/10.4269/ajtmh.15-0271
5. Coronavirus: COVID-19 Is Now Officially a pandemic, WHO says, March 11, 202012:30 PM ET, The corona virus crisis. https://www.npr.org/sections/goatsandsoda/2020/03/11/814474930/coronavirus-covid-19-is-now-officially-a-pandemic-who-says
6. How Coronavirus Spreads, Coronavirus Disease 2019 (COVID-19), Centre for Disease Control and Prevention (CDC). https://www.cdc.gov/coronavirus/2019-ncov/prevent-gettingsick/how-covidspreads.html?CDC_AA_refVal=https%3A%2F%2Fhttps://www.cdc.gov%2Fcoronavirus%2F2019-ncov%2Fprepare%2Ftransmission.html
7. Mora M et al (2006) Microbial genomes and vaccine design: refinements to the classical reversevaccinology approach. Curr Open Microbiol 9(5):532–536
8. Altschul SF, Gish W, Miller W, Myers EW, Lipman DJ (1990) Basic local alignment search tool. J Mol Biol 215:403–410. BLAST PROGRAMS
9. Gasteiger E, Gattiker A, Hoogland C, Ivanyi I, Appel RD, Bairoch A (2003) ExPASy: the proteomics server for in-depth protein knowledge and analysis. Nucleic Acids Res 31:3784–3788
10. Garcia-Boronat M, Diez-Rivero CM, Reinherz EL, Reche PA (2008) PVS: a web server for protein sequence variability analysis tuned to facilitate conserved epitope discovery. Nucleic Acids Res 36(Web Server issue):W35–W41. https://doi.org/10.1093/nar/gkn211
11. Rice P, Longden I, Bleasby A (2000) EMBOSS: The European molecular biology open software suite. Trends Genet 16(6):276–277

12. Book: Combined quantum mechanical and molecular mechanical methods (American Chemical Society ACS Symposium Series) 1997. ISBN-10: 0841235902)
13. Gasteiger E, Hoogland C, Gattiker A, Duvaud S, Wilkins MR, Appel RD, Bairoch A (2005) Protein identification and analysis tools on the ExPASy server. In: Walker JM (ed) The proteomics protocols handbook, Humana Press, pp 571–607

Chapter 10
COVID19 Rapid Diagnostic Kits

Apart from the treatment options, protocols and preventive measures against COVID, an essential step prior to the treatment is its quick and accurate diagnosis. There are several diagnostic protocols prevailing currently for the identification of COVID infection. The Gold standard available diagnosis is based on the identification of pathogen by RTPCR analysis [1]. This involves extraction of the genetic material of the virus followed by its amplification using specific primers and identification. This is the most sensitive and accurate protocol but involves high expertise and time. Thus the research was focused towards identification of Rapid testing kits which requires less time and not much expertise in handling the procedure. It is a simple KIT based method which involves the identification of specific antibodies in the blood of the tested individual against the antigens of Novel Corona Virus. The results of these tests are not as accurate as the PCR approach, yet it can be a good solution for testing huge population size and emergencies.

As of 6 March 2020, WHO listed the development laboratories and protocols for detection of virus as follows (Table 10.1).

10.1 Principle and Working of LabCorp COVID19 RT-PCR Test Kit

LabCorp COVID-19 RT-PCR test Kit is one of the most widely accepted kits because of its authenticity and validity. This kit is EUA [3] ie Accelerated Emergency Use Authorization kit. The kit is based on the RT PCR analysis of the Clinical specimens that are obtained from the nasal and throat swabs. This RT PCR based test kit performs the qualitative detection of nucleic acids from the upper and lower respiratory specimen of SARS CoV 2. Some of the specimen used for testing includes nasopharyngeal sample, oropharyngeal swabs, aspirates from lower respiratory tract, sputum, bronchoalveolar lavage and nasopharyngeal aspirate.

A. Kumar et al., *Novel Coronavirus 2019*,
SpringerBriefs in Forensic and Medical Bioinformatics,
https://doi.org/10.1007/978-981-15-7918-9_10

Table 10.1 List of different countries involved in Kits' development based on various gene targets [2]

Country	Institute	Gene targets
China	China CDC	ORF1ab and Nucleoprotein (N)
Germany	Charité	RdRP, E, N
Hong Kong	HKU	ORF1b-nsp14, N
Japan	NIID	Pancorona and multiple targets, Spike protein (Peplomer)
Thailand	National Institute of Health	N
United States	US CDC	Three targets in N gene
France	Pasteur Institute	Two targets in RdRP

The test conducted by this kit involves the use of 3 primers and their probes to identify three different regions in the SARS CoV genome which is the Nucleocapsid N gene. One of the three sets is used to identify the Human RNase P or RP in the clinical specimen. The initial step in the protocol is the isolation of the genomic RNA from the pathogen within the specimen. This is followed by its purification and amplification. Applied Biosystems QuantStudio7 Flex (QS7) instrument with 1.3 version of the software is used for the amplification of the sample. During the PCR cycle there is an annealing step where the probe binds to the specific region in the target sequence between both forward and reverse primers. The bound probe is later degraded by the Taq Polymerase enzyme during the extension phase of the PCR cycle. This reaction can be detected using the reporter dye FAM which gets separated from the quencher dye BHQI which generates a fluorescent signal. The intensity of this fluorescence is detected during each cycle using QS7. Before the interpretation of results for the patient samples all the controls have to be examined to determine the accuracy of the protocol.

The protocol involves the use of Roche MagNA Pure-96 (MP96) kit for the extraction along with the Applied Biosystems QuantStudio7 Flex (QS7) instrument with software version 1.3 (Fig. 10.1).

Some important components in the kit are as follows

(1) To eliminate the chances of errors due to sample contamination, a negative control which is molecular level nuclease free water is used in each plate of the assay.

(2) A positive template of COVID-19 that contains in vitro transcribed pure viral targeted RNA with one copy each of N1, N2 and N3 is added during the addition

Reagent	Manufacturer	Catalog #
DNA and Viral Small Volume Kit (3x192 purifications)	Roche	06543588001
TaqPath 1-Step RT-PCR Master Mix, GC (2000 reactions)	ThermoFisher	A15300
COVID-19_N1-F Primer	IDT	Custom
COVID-19_N1-R Primer	IDT	Custom
COVID-19_N1-P Probe	IDT	Custom
COVID-19_N2-F Primer	IDT	Custom
COVID-19_N2-R Primer	IDT	Custom
COVID-19_N2-P Probe	IDT	Custom
COVID-19_N3-F Primer	IDT	Custom
COVID-19_N3-R Primer	IDT	Custom
COVID-19_N3-P Probe	IDT	Custom
RP-F Primer	IDT	Custom
RP-R Primer	IDT	Custom
RP-P Probe	IDT	Custom
COVID-19_N_Positive Control	IDT	Custom
Hs_RPP30_Internal Extraction Control	IDT	Custom

Fig. 10.1   Details of reagents used in the LabCorp kit

of master mix at a concentration of copies/uL. This is also added to each of the assay plates.

(3)   To confirm the presence of nucleic acid in the sample an internal Hs_RPP30 control targeting RNase P is used. This is an extraction control.

(4)   An NEC (Negative extraction control) obtained from previous testing is used for the negative patient sample confirmation (Fig. 10.2).

10.2   RAPID Test Kit for COVID Detection Based on Antibodies Identification

RAPID test kits involve the identification of antibodies in the blood or other specimen samples based on their binding to the test antigens added in the test. The tests are mostly based on the lateral flow qualitative immune-chromatographic assay [4]. The results are read based on the development of a colored line similar to the HCG detection kit. The targeted antibodies are Immunoglobulin M-IgM, and Immunoglobulin G-IgG in the samples like blood, serum or plasma. The results in this test can be visually analyzed and obtained within a time frame of 10 min. Best approach in emergencies and is a cost effective method for diagnosis.

Most of the available RAPID test kits available globally are designed with a sample pad and a chromatographic membrane, mostly a nitrocellulose membrane that is coated with mouse anti-human Immunoglobulin M monoclonal antibody and a mouse anti-human IgG monoclonal antibody and goat anti-mouse IgG antibody. It also has a colloidal binding pad coated with labeled recombinant novel coronavirus (COVID-19) antigen and mouse IgG antibody, a liner and an absorbent pad [5]. The results are detected based on the development of colored lines on the chromatographic strip. Unfortunately, Rapid testing kits against COVID 19 has not been a great success

COVID-19 RT-PCR test results interpretation

SARS-CoV-2 N1	SARS-CoV-2 N2	SARS-CoV-2 N3	RNase P	Result Interpretation	Report	Actions
+	+	+	+/-	SARS-CoV-2 Detected	POSITIVE	Report results to sender and appropriate public health authorities.
If only one or both targets are positive			+/-	+/-	SARS-CoV-2 Detected	POSITIVE	Report results to sender and appropriate health authorities.
-	-	+	+/-	SARS-CoV-2 is Presumptive Positive	PRESUMPTIVE POSITIVE	Sample is repeated once. If the repeated result remains "PRESUMPTIVE POSITIVE", additional confirmatory testing may be conducted, if it is necessary to differentiate between SARS-CoV-2 and other SARS-like viruses for epidemiological purposes or clinical management.
-	-	-	+	SARS-CoV-2 Not Detected	NEGATIVE	Report results to sender.
-	-	-	-	Invalid Result	INVALID	Sample is repeated once. I f a second failure occurs, it is reported to sender as invalid and recommend recollection if patient is still clinically indicated.

Fig. 10.2 LabCorp COVID-19 RT-PCR test EUA Summary. Table reference: https://www.fda.gov/media/136151/download

as of now, higher percentage of false positive and false negative results have been confusing the diagnosis. It is not considered a standard approach, Governments and Authorities have been recommending getting the RT PCR testing done even if someone has tested negative with Rapid kits in COVID Cases.

References

1. van Kasteren PB et al (2020) Comparison of seven commercial RT-PCR diagnostic kits for COVID-19. J Clin Virol 128:104412
2. Coronavirus Disease 2019 (COVID-19) Novel coronavirus (2019-nCoV) real-time RT-PCR primers and probes. Centre for Disease Control and Prevention. https://www.cdc.gov/coronavirus/2019-ncov/lab/rt-pcr-panel-primer-probes.html
3. https://www.fda.gov/media/136151/download
4. https://www.biopanda.co.uk/php/products/rapid/infectious_diseases/covid19.php
5. COVID-19 IgG/IgM Rapid Test Kit, Elabscience. https://www.elabscience.com/p-covid_19_igg_igm_rapid_test-375335.html

Printed in the United States
By Bookmasters